ENVIRONMENTAL HEALTH - PHYSICAL, CHEMICAL
AND BIOLOGICAL FACTORS

U.S. WETLANDS

BACKGROUND, ISSUES AND MAJOR COURT RULINGS

ENVIRONMENTAL HEALTH - PHYSICAL, CHEMICAL AND BIOLOGICAL FACTORS

Additional books in this series can be found on Nova's website under the Series tab.

Additional e-books in this series can be found on Nova's website under the e-book tab.

ENVIRONMENTAL HEALTH - PHYSICAL, CHEMICAL
AND BIOLOGICAL FACTORS

U.S. WETLANDS

BACKGROUND, ISSUES AND MAJOR COURT RULINGS

HARRIET M. HUTSON
EDITOR

New York

Copyright © 2014 by Nova Science Publishers, Inc.

All rights reserved. No part of this book may be reproduced, stored in a retrieval system or transmitted in any form or by any means: electronic, electrostatic, magnetic, tape, mechanical photocopying, recording or otherwise without the written permission of the Publisher.

For permission to use material from this book please contact us:
Telephone 631-231-7269; Fax 631-231-8175
Web Site: http://www.novapublishers.com

NOTICE TO THE READER

The Publisher has taken reasonable care in the preparation of this book, but makes no expressed or implied warranty of any kind and assumes no responsibility for any errors or omissions. No liability is assumed for incidental or consequential damages in connection with or arising out of information contained in this book. The Publisher shall not be liable for any special, consequential, or exemplary damages resulting, in whole or in part, from the readers' use of, or reliance upon, this material. Any parts of this book based on government reports are so indicated and copyright is claimed for those parts to the extent applicable to compilations of such works.

Independent verification should be sought for any data, advice or recommendations contained in this book. In addition, no responsibility is assumed by the publisher for any injury and/or damage to persons or property arising from any methods, products, instructions, ideas or otherwise contained in this publication.

This publication is designed to provide accurate and authoritative information with regard to the subject matter covered herein. It is sold with the clear understanding that the Publisher is not engaged in rendering legal or any other professional services. If legal or any other expert assistance is required, the services of a competent person should be sought. FROM A DECLARATION OF PARTICIPANTS JOINTLY ADOPTED BY A COMMITTEE OF THE AMERICAN BAR ASSOCIATION AND A COMMITTEE OF PUBLISHERS.

Additional color graphics may be available in the e-book version of this book.

Library of Congress Cataloging-in-Publication Data

ISBN: 978-1-63117-800-9

Published by Nova Science Publishers, Inc. † New York

CONTENTS

Preface		vii
Chapter 1	Wetlands: An Overview of Issues *Claudia Copeland*	1
Chapter	The Wetlands Coverage of the Clean Water Act (CWA): Rapanos and Beyond *Robert Meltz and Claudia Copeland*	41
Chapter 3	The Supreme Court Addresses Corps of Engineers Jurisdiction Over "Isolated Waters": The SWANCC Decision *Robert Meltz and Claudia Copeland*	77
Chapter 4	Clean Water Act: A Summary of the Law *Claudia Copeland*	93
Index		109

PREFACE

Wetlands, with a variety of physical characteristics, are found throughout the country. They are known in different regions as swamps, marshes, fens, potholes, playa lakes, or bogs. Although these places can differ greatly, they all have distinctive plant and animal assemblages because of the wetness of the soil. Some wetland areas may be continuously inundated by water, while other areas may not be flooded at all. In coastal areas, flooding may occur daily as tides rise and fall. Prior to the mid-1980s, federal laws and policies to protect wetlands were generally limited to providing habitat for migratory waterfowl, especially ducks and geese. Some laws encouraged destruction of wetland areas, including selected provisions in the federal tax code, public works legislation, and farm programs. Since the mid-1980s, the values of wetlands have been recognized in different ways in numerous national policies, and federal laws either encourage wetland protection, or prohibit or do not support their destruction. This book discusses wetlands and the Clean Water Act (CWA) in the United States.

Chapter 1 – Recent Congresses have considered numerous policy topics that involve wetlands. Many reflect issues of long-standing interest, such as applying federal regulations on private lands, wetland loss rates, and restoration and creation accomplishments. The issue receiving the greatest attention recently has been determining which wetlands should be included and excluded from permit requirements under the Clean Water Act's (CWA) program that regulates waste discharges affecting wetlands, which is administered by the Army Corps of Engineers and the Environmental Protection Agency (EPA). As a result of Supreme Court rulings in 2001 (in the *SWANCC* case) that narrowed federal regulatory jurisdiction over certain isolated wetlands, and in 2006 (in the *Rapanos-Carabell* decision), the jurisdictional reach of the permit program has been narrowed. In response, legislation intended to reverse the Court's rulings in these cases has been introduced regularly since the 107th Congress. In the 111th Congress, for the first time, one such bill was approved by a congressional committee (S. 787, the Clean Water Restoration Act); no further legislative action occurred. The Obama Administration did not endorse any specific legislation, but identified general principles for legislation that would clarify waters protected by the CWA. In 2011 the Administration proposed new interpretive guidance intended to clarify jurisdictional uncertainties resulting from the Court's rulings and to apply protection to additional waters and wetlands, a conclusion that pleased some observers and alarmed others. In September 2013, EPA and the Corps withdrew the 2011 proposed guidance, which had not been finalized, in favor of draft revised regulations, which are being reviewed by the Office of Management and Budget. Wetland protection efforts continue to engender controversy over issues of science and policy. Controversial topics include the rate and pattern of loss, whether all wetlands should be protected in a single fashion, the effectiveness of the current suite of laws in protecting them, and the fact that 75% of remaining U.S. wetlands are located on private lands. Many public and private efforts have sought to mitigate damage to wetlands and to protect them through acquisition, restoration, enhancement, and creation, particularly coastal wetlands. While recent data indicate success in some restoration efforts, leading to increases in some types of wetlands, many scientists question if restored or created wetlands provide equivalent replacement for natural wetlands that contribute multiple environmental services and values. One reason for controversies about wetlands is that they occur in a wide variety of physical forms, and the numerous values they provide, such as wildlife habitat, also vary widely. In addition, the total wetland acreage in the lower 48 states is estimated to have declined from more than 220 million acres three centuries ago to 110.1 million acres in 2009. The national policy goal of no net loss, endorsed by administrations for the past two decades, had been reached by 2004, according to the Fish and Wildlife Service, as the rate of loss had been more than offset by net gains through expanded restoration efforts authorized in multiple laws. However, more recent data show wetlands losses of nearly 14,000 acres per year. Many protection advocates say that gains do not necessarily account for the changes in quality of the remaining wetlands, and many also view federal protection efforts as inadequate or uncoordinated. Others, who advocate the rights of property owners and development interests, characterize these efforts as too intrusive. Numerous state and local wetland programs add to the complexity of the protection effort.

Chapter 2 – In 1985 and 2001, the Supreme Court grappled with issues as to the geographic scope of the wetlands permitting program in the federal Clean Water Act (CWA). In 2006, the Supreme Court rendered a third decision, *Rapanos v. United States*, on appeal from two Sixth Circuit rulings. The Sixth Circuit rulings offered the Court a chance to clarify the reach of CWA jurisdiction over wetlands adjacent only to *non*navigable tributaries of traditional navigable waters—including tributaries such as drainage ditches and canals that may flow intermittently. (Jurisdiction over wetlands adjacent to traditional navigable waters was established in the 1985 decision). The Court's decision provided little clarification, however, splitting 4-1-4. The four-Justice plurality decision, by Justice Scalia, said that the CWA covers only wetlands connected to relatively permanent bodies of water (streams, rivers, lakes) by a continuous surface connection. Justice Kennedy, writing alone, demanded a substantial nexus between the wetland and a traditional navigable water, using an ambiguous ecological test. Justice Stevens, for the four dissenters, would have upheld the existing broad reach of Corps of Engineers/EPA regulations. Because no rationale commanded the support of a majority of the Justices, lower courts are extracting different rules of decision from *Rapanos* for resolving future cases. Corps/EPA guidance issued in December 2008 says that a wetland generally is jurisdictional if it satisfies either the plurality or Kennedy tests. In April 2011, the agencies proposed revised guidance intended to clarify whether waters are protected by the CWA, but this proposal was controversial. The ambiguity of the *Rapanos* decision and questions about the agencies' guidance have increased pressure on EPA and the Corps to initiate a rulemaking to promulgate new regulations. In September 2013, EPA and the Corps withdrew the controversial proposed guidance and submitted a draft rule to the Office of Management and Budget for review. The substance of the draft rule, and when it might be proposed, are unknown for now. There also has been pressure on Congress to provide legislative clarification. In the 111th Congress, legislation intended to do so was approved by a Senate committee, but no further legislative action occurred. Similar legislation was not introduced in the 112th Congress or so far in the 113th Congress. Instead, proposals to bar issuance of the Corps/EPA revised guidance and to narrow the regulatory scope of the CWA have been introduced. The legal and policy questions associated with *Rapanos*—regarding the outer geographic limit of CWA jurisdiction and the consequences of restricting that scope—have challenged regulators, landowners and developers, and policymakers for 40 years. The answer may determine the reach of CWA regulatory authority not only for the wetlands permitting program but also for other CWA programs, since the CWA uses but one jurisdiction-defining phrase ("navigable waters") throughout the statute. While regulators and the regulated community debate the legal dimensions of federal jurisdiction under the CWA, scientists contend that there are no discrete, scientifically supportable boundaries or criteria along the continuum of wetlands to separate them into meaningful ecological or hydrological compartments. Wetland scientists believe that all such waters are critical for protecting the integrity of waters, habitat, and wildlife downstream. Changes in the limits of federal jurisdiction highlight the role of states in protecting waters not addressed by federal law. From the states' perspective, federal programs provide a baseline for consistent, minimum standards to regulate wetlands and other waters. Most states are either reluctant or unable to take steps to protect non-jurisdictional waters through legislative or administrative action.

Chapter 3 – On January 9, 2001, the Supreme Court handed down *Solid Waste Agency of Northern Cook County (SWANCC) v. U.S. Army Corps of Engineers*. At issue in *SWANCC* was the scope of Clean Water Act section 404, which requires permits for the discharge of dredged or fill materials into "navigable waters," defined by the Act as "waters of the United States." Section 404 is the charter for the federal wetlands permitting program. *SWANCC* explicitly held that the Corps of Engineers' use of the "migratory bird rule," adopted by the agency to interpret the reach of its section 404 authority over "isolated waters" (including isolated wetlands), exceeded the authority granted by that section. Looking at the decision's rationale rather than its holding, however, *SWANCC* may be read more broadly to bar assertion of section 404 jurisdiction over isolated waters on *any* basis, migratory bird rule or otherwise. The migratory bird rule asserted that section 404 covers, among other waterbodies, isolated waters "which are or would be used as habitat by ... migratory birds that cross state lines ...." In 1985, the Supreme Court had sustained the assertion by the Corps and EPA that waters and wetlands *adjacent to* navigable waters, interstate waters, or their tributaries are "waters of the United States" under section 404. The question left for *SWANCC* was whether waters and wetlands not so adjacent – "isolated waters" – also are so covered. Such jurisdictional lines stand in contrast to the scientists' perspective, which recognizes the value of wetlands based on water quality and other physical functions which they perform, irrespective of whether the wetlands are isolated or contiguous to other waters. Estimates of waters and wetland acreage likely to be removed from the section 404 permitting program as a result of the *SWANCC* decision are very difficult to assess, in part because of questions about Corps and EPA interpretation of the ruling, but the decision may affect up to 79% of wetland acreage. One likely result is that in those cases where case-by-case evaluations will be required to determine if regulatory jurisdiction exists, the length of time to obtain section 404 permits will be longer than in the past. If federal jurisdiction is diminished, the responsibility to protect affected wetlands falls on states and local governments. A comprehensive picture of their ability to protect wetlands, under various possible state and local authorities, is difficult to draw together. Whether states will act to fill in the gap left by removal of some federal jurisdiction through new laws or programs raises difficult political and resource questions. The *SWANCC* decision also raises issues for Congress. First is whether confusion that may now exist about the extent of Clean Water Act jurisdictional waters and wetlands should be resolved, and what constitutional limits may apply. Second is whether to provide federal resources and incentives to encourage expansion of state wetlands protection and regulatory programs or others that encourage acquisition and conservation of wetlands.

Chapter 4 – The principal law governing pollution of the nation's surface waters is the Federal Water Pollution Control Act, or Clean Water Act. Originally enacted in 1948, it was totally revised by amendments in 1972 that gave the act its current dimensions. The 1972 legislation spelled out ambitious programs for water quality improvement that have since been expanded and are still being implemented by industries and municipalities. This report presents a summary of the law, describing the statute without discussing its implementation. The Clean Water Act consists of two major parts, one being the provisions which authorize federal financial assistance for municipal sewage treatment plant construction. The other is the regulatory requirements that apply to industrial and municipal dischargers. The act has been termed a technology-forcing statute because of the rigorous demands placed on those who are regulated by it to achieve higher and higher levels of pollution abatement under deadlines specified in the law. Early on, emphasis was on controlling discharges of conventional pollutants (e.g., suspended solids or bacteria that are biodegradable and occur naturally in the aquatic environment), while control of toxic pollutant discharges has been a key focus of water quality programs more recently. Prior to 1987, programs were primarily directed at point source pollution, that is, wastes discharged from discrete sources such as pipes and outfalls. Amendments to the law in that year authorized measures to address nonpoint source pollution (runoff from farm lands, forests, construction sites, and urban areas), which is estimated to represent more than 50% of the nation's remaining water pollution problems.

The act also prohibits discharge of oil and hazardous substances into U.S. waters. Under this act, federal jurisdiction is broad, particularly regarding establishment of national standards or effluent limitations. Certain responsibilities are delegated to the states, and the act embodies a philosophy of federal-state partnership in which the federal government sets the agenda and standards for pollution abatement, while states carry out day-to-day activities of implementation and enforcement. To achieve its objectives, the act is based on the concept that all discharges into the nation's waters are unlawful, unless specifically authorized by a permit, which is the act's principal enforcement tool. The law has civil, criminal, and administrative enforcement provisions and also permits citizen suit enforcement. Financial assistance for constructing municipal sewage treatment plants and certain other types of water quality improvements projects is authorized under title VI. It authorizes grants to capitalize State Water Pollution Control Revolving Funds, or loan programs. States contribute matching funds, and under the revolving loan fund concept, monies used for wastewater treatment construction are repaid to states, to be available for future construction in other communities.

In: U.S. Wetlands
Editor: Harriet M. Hutson

ISBN: 978-1-63117-800-9
© 2014 Nova Science Publishers, Inc.

Chapter 1

WETLANDS: AN OVERVIEW OF ISSUES[*]

Claudia Copeland

SUMMARY

Recent Congresses have considered numerous policy topics that involve wetlands. Many reflect issues of long-standing interest, such as applying federal regulations on private lands, wetland loss rates, and restoration and creation accomplishments.

The issue receiving the greatest attention recently has been determining which wetlands should be included and excluded from permit requirements under the Clean Water Act's (CWA) program that regulates waste discharges affecting wetlands, which is administered by the Army Corps of Engineers and the Environmental Protection Agency (EPA). As a result of Supreme Court rulings in 2001 (in the *SWANCC* case) that narrowed federal regulatory jurisdiction over certain isolated wetlands, and in 2006 (in the *Rapanos-Carabell* decision), the jurisdictional reach of the permit program has been narrowed. In response, legislation intended to reverse the Court's rulings in these cases has been introduced regularly since the 107th Congress. In the 111th Congress, for the first time, one such bill was approved by a congressional committee (S. 787, the Clean Water Restoration Act); no further legislative action occurred. The Obama Administration did not endorse any specific legislation, but identified general principles for legislation that would clarify waters protected by the CWA. In 2011 the

---

[*] This is an edited, reformatted and augmented version of Congressional Research Service Publication, No. RL33483, dated December 5, 2013.

Administration proposed new interpretive guidance intended to clarify jurisdictional uncertainties resulting from the Court's rulings and to apply protection to additional waters and wetlands, a conclusion that pleased some observers and alarmed others. In September 2013, EPA and the Corps withdrew the 2011 proposed guidance, which had not been finalized, in favor of draft revised regulations, which are being reviewed by the Office of Management and Budget.

Wetland protection efforts continue to engender controversy over issues of science and policy. Controversial topics include the rate and pattern of loss, whether all wetlands should be protected in a single fashion, the effectiveness of the current suite of laws in protecting them, and the fact that 75% of remaining U.S. wetlands are located on private lands.

Many public and private efforts have sought to mitigate damage to wetlands and to protect them through acquisition, restoration, enhancement, and creation, particularly coastal wetlands. While recent data indicate success in some restoration efforts, leading to increases in some types of wetlands, many scientists question if restored or created wetlands provide equivalent replacement for natural wetlands that contribute multiple environmental services and values.

One reason for controversies about wetlands is that they occur in a wide variety of physical forms, and the numerous values they provide, such as wildlife habitat, also vary widely. In addition, the total wetland acreage in the lower 48 states is estimated to have declined from more than 220 million acres three centuries ago to 110.1 million acres in 2009. The national policy goal of no net loss, endorsed by administrations for the past two decades, had been reached by 2004, according to the Fish and Wildlife Service, as the rate of loss had been more than offset by net gains through expanded restoration efforts authorized in multiple laws. However, more recent data show wetlands losses of nearly 14,000 acres per year. Many protection advocates say that gains do not necessarily account for the changes in quality of the remaining wetlands, and many also view federal protection efforts as inadequate or uncoordinated. Others, who advocate the rights of property owners and development interests, characterize these efforts as too intrusive. Numerous state and local wetland programs add to the complexity of the protection effort.

INTRODUCTION

Wetlands, with a variety of physical characteristics, are found throughout the country. They are known in different regions as swamps, marshes, fens, potholes, playa lakes, or bogs. Although these places can differ greatly, they all have distinctive plant and animal assemblages because of the wetness of the soil. Some wetland areas may be continuously inundated by water, while other areas may not be flooded at all. In coastal areas, flooding may occur daily as tides rise and fall.

Prior to the mid-1980s, federal laws and policies to protect wetlands were generally limited to providing habitat for migratory waterfowl, especially ducks and geese. Some laws encouraged destruction of wetland areas, including selected provisions in the federal tax code, public works legislation, and farm programs.

Since the mid-1980s, the values of wetlands have been recognized in different ways in numerous national policies, and federal laws either encourage wetland protection, or prohibit or do not support their destruction. These laws, however, do not add up to a fully consistent or comprehensive national approach. The central federal regulatory program, found in Section 404 of the Clean Water Act, requires permits for the discharge of dredged or fill materials into many but not all wetland areas. However, other activities that may adversely affect wetlands do not require permits, and some places that scientists define as wetlands are exempt from this permit program because of physical characteristics or the type of activity that takes place. One agricultural program, swampbuster, is a disincentive program that indirectly protects wetlands by making farmers who drain wetlands ineligible for federal farm program benefits; those who do not receive these benefits (62% of all farmers received no direct payments from the farm subsidy program in 2007) have no reason to observe the requirements of this program. Numerous other acquisition, protection, and restoration programs complete the current federal effort.

Although numerous wetland protection bills have been introduced in recent Congresses, the most significant new wetlands legislation to be enacted has been in farm bills, in 1996, 2002, and 2008. During this period, Congress also reauthorized several wetlands programs, mostly setting higher appropriations ceilings, without making significant shifts in policy. The George W. Bush Administration endorsed wetland protection in legislation, such as the farm bill and the North American Wetlands Conservation Act reauthorization, and at events, such as Earth Day presentations. The Bush Administration also issued rules on mitigation policies. The Obama Administration has proposed controversial guidance on regulatory program jurisdiction (see discussion below).

Congress has provided a forum in numerous hearings where conflicting interests in wetland issues have been debated. These debates encompass disparate scientific and programmatic questions, and conflicting views of the role of government where private property is involved. Broadly speaking, the conflicts are between:

- Environmental interests and wetland protection advocates who have been pressing for greater wetlands protection as multiple values have been more widely recognized, by improving coordination and consistency among agencies and levels of governments, and strengthened programs; and
- Others, including landowners, farmers, and small businessmen, who counter that protection efforts have gone too far, by aggressively regulating privately owned wet areas that provide few wetland values. They have been especially critical of the U.S. Army Corps of Engineers (Corps) and the U.S. Environmental Protection Agency (EPA), asserting that they administer the Section 404 program in an overzealous and inflexible manner.

Wetland legislative activity in the 110th Congress centered broadly on two issues. One was on wetlands conservation provisions in the 2008 farm bill, which was enacted in June 2008 (Food, Conservation, and Energy Act of 2008, P.L. 110-246). The new law reauthorized and increased the acreage enrollment cap in the wetlands reserve program, with a goal of enrolling 250,000 acres annually, and extended provisions to enroll up to a million acres of wetlands and buffers in the Conservation Reserve Program. Other agricultural conservation programs, while lacking explicit wetlands protection provisions, are still likely to be beneficial to wetlands.

The second major area of recent legislative interest was proposals to reverse Supreme Court rulings that addressed and narrowed the scope of geographic jurisdiction of wetlands regulations under the Clean Water Act. This interest arises because federal courts have played a key role in interpreting and clarifying the limits of federal jurisdiction to regulate activities that affect "waters of the United States," including wetlands, especially since a 2001 Supreme Court ruling in the so-called *SWANCC* decision and another in 2006 in *Rapanos v. United States*. In the 111th Congress, legislation intended to reverse the *SWANCC* and *Rapanos* rulings was approved by a Senate Committee (S. 787), but no further action occurred. Legislation that instead would narrow the definition of "waters of the United States" also has been introduced, such as S. 890/H.R. 3377 in the 113th Congress.

WETLANDS: SCIENCE AND INFORMATION

Scientific questions about wetlands, with answers that can be important to policy makers, include how to define wetlands; how to catalogue the rate and pattern of wetland declines and losses as well as restorations and increases; and how to assess the importance of wetland changes to broader ecosystems. Wetlands science has made considerable strides in developing a fuller and more sophisticated knowledge about many aspects of wetlands in the more than two decades since protecting wetlands became a general policy goal in federal law and program administration.[1]

Two topics where scientific information and wetland protection policies remain inconsistent continue to be: should all regulated wetlands be treated equally; and if all scientifically defined wetlands are not covered by the federal regulatory program, what subset should be covered, and how should such decisions be made? While discussion of either question has major science elements, both are primarily addressed in the section below about the Clean Water Act Section 404 program.

What Is a Wetland?

Scientists generally agree that the presence of a wetland can be determined by a combination of soils, plants, and hydrology. The only definition of wetlands in law, in the swampbuster provisions of farm legislation (P.L. 99-198) and in the Emergency Wetlands Resources Act of 1986 (P.L. 99-645), lists those three components. This definition does not include more specific criteria, such as exactly what conditions must be present and for how long, thus leaving interpretation to scientists and regulators on a case-by-case basis. Controversies are exacerbated when many sites that have those three components and are identified as wetlands by experts, either may have wetland characteristics only some portion of the time, or may not look like what many people visualize as wetlands. Also, many of these sites have been directly or indirectly modified by human activities that diminish their appearance (and their ability to perform wetland functions).

Wetlands currently subject to federal regulation are a large subset of all places that members of the scientific community would call a wetland. These regulated wetlands, under the Section 404 program discussed below, are currently identified using technical criteria in a wetland delineation manual issued by the Corps in 1987. This manual was prepared jointly and is used by all federal agencies to carry out their responsibilities under this program (the Corps, EPA, Fish and Wildlife Service (FWS), and the National Marine Fisheries Service (NMFS)). It provides guidance and field-level consistency for the agencies that have roles in wetland regulatory protection. (A second and slightly different manual, agreed to by the Corps and the Natural Resources Conservation Service (NRCS), is used for delineating wetlands on agricultural lands.) While the agencies try to improve the objectivity and consistency of wetland identification and delineation, judgment continues to play a role and can lead to site-specific controversies. Cases discussed below (see "Section 404 Judicial Proceedings: *SWANCC* and *Rapanos*") center on whether wetlands should be included or exempted from the regulatory program in certain circumstances, such as the physical setting.

What Functional Values Are Provided by Wetlands?

Functional values, both ecological and economic, at each wetland depend on its location, size, and relationship to adjacent land and water areas. Many of these values have been recognized only recently. Historically, many federal programs encouraged wetlands to be drained or altered because they were seen as having little value as wetlands (for example, flood protection programs of the Corps and Department of Agriculture have modified or eliminated many floodplain wetlands through alterations of the hydraulic/hydrologic regime). Wetland values can include

- habitat for aquatic birds and other animals and plants, including numerous threatened and endangered species; production of fish and shellfish;
- water storage, including mitigating the effects of floods and droughts;
- water purification;
- recreation;
- timber production;
- food production;
- education and research; and
- open space and aesthetic values.

Usually wetlands provide some combination of these values; single wetlands rarely provide all of these values. The composite value typically declines when wetlands are altered. In addition, the effects of alteration often extend well beyond the immediate area, because wetlands are usually part of a larger water system. For example, conversion of wetlands to urban uses has increased flood damages; this value has received considerable attention as the costs of natural disaster costs mounted since the 1990s.

How Fast Are Wetlands Disappearing, and How Many Acres Are Left?

A number of reports document changes in wetland acres. The U.S. Fish and Wildlife Service (FWS) periodically surveys national net trends in wetland acreage using the National Wetlands Inventory (NWI). It has estimated that when European settlers first arrived, wetland acreage in the area that would become the 48 states was more than 220 million acres, or about 5% of the total land area. According to its most recent report of national trends, issued in 2011, total wetland acreage in 2009 was estimated to be 110.1 million acres.[2] Until recently, NWI data had shown small annual gains overall in wetland acreage. However, the 2009 total was a slight decline in acreage over the previous five years (62,300 acres), or about 13,800 acres lost per year, reflecting a combination of some losses and some gains in acres and types of wetlands across the country. FWS also has published reports on wetland status and trends in several individual regions and states, such as Florida, Texas, Delaware, South Carolina, and Alaska.[3]

Of particular interest to scientists and natural resource managers are coastal wetlands, which provide important ecosystems services, because they serve as buffers to protect coastal areas from storm damage and sea level rise, while providing habitat for fish, shellfish, and wildlife that are commercially and recreationally important. Coastal watersheds, where these wetlands are located, are affected by population growth more than non-coastal areas, since 52% of the total U.S. population lives in counties that drain to coastal watersheds, although these counties are less than 20% of U.S. land area, excluding Alaska. Coastal wetlands are vulnerable to direct and indirect effects of residential and commercial development, pollutant discharges, and other human activities. A 2013 report by the FWS and National Oceanic and Atmospheric Administration (NOAA) found that in 2009 there were an estimated 41.1 million acres of wetlands in the coastal watersheds of the United States, representing 37.3% of total wetland area in the lower 48 states. The report also found that U.S. coastal wetlands are vanishing at a rate of more than 80,000 acres per year, about six times greater than the estimated rate of wetland loss for the entire United States. The increased loss, measured between 2004 and 2009, was attributed to severe weather in the Gulf of Mexico and urban and rural development in other areas, and the reported loss was 25% greater than the annual loss rate found in a previous report covering the years 1998 to 2004. The largest loss, according to the report, was in the Gulf region, where 257,150 acres of coastal wetlands disappeared due to erosion and/or inundation. Throughout the Gulf region, saltwater wetlands have been adversely affected by the cumulative effect of oil and gas development that increased their vulnerability to intense storms.[4]

A study of national wetland condition by the EPA together with states, tribes, and other federal partners is underway and is expected to provide information on both quality and quantity of wetlands in the United States.[5]

In 2002, the George W. Bush Administration endorsed the concept of "no-net-loss" of wetlands— a goal declared by President George H. W. Bush in 1988 and also embraced by President Clinton to balance wetlands losses and gains in the short term and achieve net gains in the long term. On Earth Day 2004, President Bush announced a new national goal, moving beyond no-net-loss to achieve an overall increase of wetlands. The goal was to create, improve, and protect at least 3 million wetland acres over the next five years in order to increase overall wetland acres and quality. (By comparison, the Clinton Administration in 1998 announced policies intended to achieve overall wetland increases of 200,000 acres per year by 2005.) The Bush strategy also called for better tracking of wetland programs and enhanced local and private sector collaboration.

In April 2008, the Bush Administration issued a report saying that more than 3.6 million acres of wetlands had been restored, protected, or improved as part of the President's program to create, improve and protect wetlands, and that the number was expected to climb to 4.5 million acres by the original date set by that program—Earth Day 2009.[6] The report documents gains, but not offsetting loses. It summarizes accomplishments for each federal wetland conservation program. Environmental groups criticized the report as presenting an incomplete picture, because it fails to mention wetlands lost to agriculture and development.

Numerous shifts in federal policies since 1985 (and changes in economic conditions as well) strongly influence wetland loss patterns, but the composite effects remain unmeasured beyond these raw numbers. There usually is a large time lag between the announcement and implementation of changes in policy, and collection and release of data that measure how these changes affect loss rates. Also, it is often very difficult to distinguish the role that policy changes play from other factors, such as agricultural markets, development pressures, and land markets.

Further, these data only measure acres. This may have been appropriate two or three decades ago when scientists knew less about how to measure the specific functions and values found in wetlands. By providing data limited to number of acres, these data provide few insights into changes in their quality, as measured by the values they provide, which is often determined by factors such as where a wetland is located in a watershed, and what are the surrounding land uses. Scientists caution that there are a number of questions about the qualitative and ecological integrity of existing wetlands. The wetlands trends data reported by FWS in 2011 show increases in certain types of freshwater wetlands since 2004, particularly freshwater ponds constructed to replace lost wetlands. However, FWS noted that there is no clear scientific consensus about the functional equivalency of replacement wetlands.

Wetlands and Climate Change

As described above, coastal wetlands provide critical services such as absorbing energy from coastal storms, preserving shorelines, protecting human populations and infrastructure, absorbing pollutants, and serving as critical habitat for migratory species. Many scientists believe that these resources and services will be threatened as sea-level rise associated with a changing climate inundates wetlands. Due in part to their limited capacity for adaptation, wetlands have been considered among the ecosystems most vulnerable to a changing climate. Changes in climatic conditions that affect water conditions (e.g., wetter, drier, more saline) are predicted to have substantial impact on species that use wetlands and on ecosystem services provided by wetlands, or make efforts to reestablish wetlands more challenging.[7]

In 2010, a group of international scientists published results of research modeling efforts to identify conditions under which coastal wetlands could survive rising sea level. Using a rapid sea-level rise scenario, the scientists estimated that most coastal wetlands worldwide will experience inundation that leads to rapid and irreversible conversion of marshland into unvegetated, subtidal surfaces and will disappear near the end of the 21st century. Under moderate and slow sea-level rise scenarios, some coastal wetlands would be vulnerable to inundation, depending on amounts of sediment present: larger amount of sediment would enable the wetland to adapt and modify naturally and thus be more likely to survive sea-level rise.[8]

Coastal wetlands also serve as a "sink" for absorbing carbon dioxide (CO_2), the most common greenhouse gas (GHG) that is associated with climate change. Scientists recognize that tidal wetlands hold large amounts of carbon, some within standing plant biomass, but most within deep organic-bearing soils. Carbon that is stored in soils has been built up over millennia and reflects pools of CO_2 that have been transferred from the atmosphere and sequestered within roots and other organic material.[9] However, the loss of wetland areas, for example through inundation and erosion, eliminates its ongoing sequestration capacity, and draining wetlands for development releases within a few decades carbon that took centuries to accumulate. In 2011, the World Bank released a report which concluded that drainage and degradation of coastal wetlands has become a major cause of carbon dioxide emissions that contribute to climate change, large enough globally that carbon dioxide emissions from drained coastal wetlands should be included in carbon accounting and emission inventories, and in policy frameworks to reduce emissions.[10] Some policymakers concerned with mitigating climate change have begun to consider whether it is possible to halt the release of carbon from converted or eroded wetlands and reverse carbon losses through wetland restoration. Further, some are considering whether the ecosystem benefits of wetlands, from a carbon sequestration standpoint, can be quantified in financial terms to enable use of wetlands restoration and management as potential generators of GHG offsets in the context of climate change policy.[11]

SELECTED FEDERAL WETLANDS PROGRAMS

Federal program issues include the administration of programs to protect, restore, or mitigate wetland resources (especially the Clean Water Act Section 404 program); relationships between agricultural and regulatory programs; whether all wetlands should be treated the same in federal programs, and which wetlands should be subject to regulation; and whether protecting wetlands by acres is an effective proxy for protecting wetlands based on the functions they perform and the values they provide. In addition, private property questions are raised, because almost three-quarters of the remaining wetlands are located on private lands. Some property owners believe that they should be compensated when federal programs limit how they can use their land, and for decisions that arguably diminish the value of the land.

The Clean Water Act Section 404 Program

The principal federal program that provides regulatory protection for wetlands is found in Section 404 of the Clean Water Act (CWA). Its intent is to protect water and adjacent wetland areas from adverse environmental effects due to discharges of dredged or fill material. Enacted in 1972, Section 404 requires landowners or developers to obtain permits from the Corps of Engineers to carry out activities involving disposal of dredged or fill materials into waters of the United States, including wetlands.

The Corps has long had regulatory jurisdiction over dredging and filling, starting with the River and Harbor Act of 1899. The Corps and EPA share responsibility for administering the Section 404 program. Other federal agencies, including NRCS, FWS, and NMFS, also have roles in this process. In the 1970s, legal decisions in key cases led the Corps to revise this program to incorporate broad jurisdictional definitions in terms of both regulated waters and adjacent wetlands. Section 404 was last amended in 1977.

This judicial/regulatory/administrative evolution of the Section 404 program has generally pleased those who view it as a critical tool in wetland protection, but dismayed others who would prefer more limited Corps jurisdiction or who see the expanded regulatory program as intruding on private land-use decisions and treating wetlands of widely varying value similarly. Underlying this debate is the more general question of whether Section 404 is the best approach to federal wetland protection.

Some wetland protection advocates have proposed that it be replaced or greatly altered. First, they point out that it governs only the discharge of dredged or fill material, while not regulating other acts that drain, flood, or otherwise reduce functional values. Second, because of exemptions provided in 1977 amendments to Section 404, major categories of activities are not required to obtain permits. These include normal, ongoing farming, ranching, and silvicultural (forestry) activities. Further, permits generally are not required for activities which drain wetlands (only for those that fill wetlands), which excludes a large number of actions with potential to alter wetlands. Third, in the view of protection advocates, the multiple values that wetlands can provide (e.g., fish and wildlife habitat, flood control) are not effectively recognized through a statutory approach based principally on water quality, despite the broad objectives of the Clean Water Act.

The Permitting Process

The Corps' regulatory process involves both general permits for actions by private landowners that are similar in nature and will likely have a minor effect on wetlands, and individual permits for more significant actions. According to the Corps, it evaluates more than 85,000 permit requests annually. Of those, more than 90% are authorized under a general permit, which can apply regionally or nationwide, and is essentially a permit by rule, meaning the proposed activity is presumed to have a minor impact, individually and cumulatively. Most general permits do not require pre-notification or prior approval by the Corps. About 9% of all permits are required to go through the more detailed evaluation for a standard individual permit, which may involve complex proposals or sensitive environmental issues and can take 180 days or longer for a decision. Less than 0.3% of permits are denied; most other individual permits are modified or conditioned before issuance. About 5% of applications are withdrawn prior to a permit decision.

Regulatory procedures on individual permits allow for interagency review and comment, a coordination process that can generate delays and an uncertain outcome, especially for environmentally controversial projects. EPA is the only federal agency having veto power over a proposed Corps permit; EPA has used its veto authority 13 times in the 30-plus years since the program began. However, critics have charged that implied threats of delay by the FWS and others practically amount to the same thing. Reforms during the Reagan, George H.W. Bush, and Clinton Administrations streamlined certain of these procedures, with the intent of speeding up and clarifying the Corps' full regulatory program, but concerns continue over both process and program goals.

Controversy also surrounded revised regulations issued by EPA and the Corps in 2002, which redefine two key terms in the 404 program: "fill material" and "discharge of fill material." These definitions are important, because material defined as "fill" is regulated and permitted under Section 404 procedures, while other waste discharges are regulated under more stringent CWA rules and procedures. The agencies said that the revisions were intended to clarify certain confusion in their joint administration of the program due to previous differences in how the two agencies defined those terms. However, environmental groups contended that the changes allow for less restrictive and inadequate regulation of certain disposal activities, including disposal of coal mining waste, which could be harmful to aquatic life in streams. Legislation to reverse the agencies' action by clarifying in the law that fill material cannot be composed of waste has been introduced regularly since the 107th Congress, including H.R. 1837 in the 113th Congress, but no further action has occurred.[12]

As previously described, three criteria—hydrology, soil type, and plants—are used in making wetlands delineations under several environmental laws and programs, including Section 404 permitting. Scientists generally agree that each of the three parameters must be met to identify an area as a wetland. Because growth of plants in wetland areas typically is contingent on the presence of hydric soils and the availability of sufficient water, the vegetation parameter often is determinative of whether an area qualifies as a wetland or not. In May 2012, the Corps revised the National Wetlands Plant List (NWPL), which is used by federal and state agencies for determining whether a particular area contains a prevalence of hydrophytic (i.e., wetland) vegetation. This is the first major revision of the plant list since its publication in 1988 and is intended to improve the accuracy of the overall list. The updated list contains 8,200 plant species, an increase of 1,472 species, or 22%, primarily as a result of new taxonomic interpretations. The Corps said it does not expect major changes to wetland delineations as a result of the updated list, but some commenters contend that the new list is likely to cause more areas to qualify as wetlands.[13]

Section 404 authorizes states to assume many of the Corps' permitting responsibilities. Two states have done this: Michigan (in 1984) and New Jersey (in 1992). Others reasons cited for not joining these two states include the complex process of assumption, the anticipated cost of running a program, and the continued involvement of federal agencies because of statutory limits on waters that states could regulate. Efforts continue to encourage more states to assume program responsibility.

EPA's Veto of Section 404 Permits

In addition to directing EPA to issue the environmental guidelines used by the Corps to evaluate permit applications (CWA Section 404(b)), Section 404 also authorizes EPA to prohibit or otherwise restrict the specification by the Corps of a site for the discharge of dredged or fill material, if the agency determines that the activity will have an unacceptable adverse effect on water supplies, fish, wildlife, or recreational areas. EPA has used this veto authority, under Section 404(c), only 13 times since 1972.

In January 2011, the agency issued the 13th veto when it vetoed a permit for a surface coal mining operation in West Virginia. According to EPA, the Spruce No. 1 mine in Logan County, West Virginia, as proposed, would be one of the largest surface mining operations ever authorized in Appalachia, and waste disposal from the mine would bury over seven miles of streams, directly impact 2,278 acres of forestland, and degrade water quality in streams adjacent to the mine. The Corps issued a permit for the project in 2007, but it was subsequently delayed by litigation and has been operating on a limited scale since then. EPA acknowledged that the project has been modified in order to reduce impacts, but the veto determination was based on the agency's conclusion that the project could result in unacceptable adverse impacts to wildlife and fishery resources.[14]

EPA's veto of the permit has been very controversial, in part because it involves cancelling a permit previously issued by the Corps.[15] Coal industry groups and organizations representing manufacturing and other sectors have been highly critical of EPA's actions, many saying that to revoke an existing permit creates huge uncertainty about whether water quality permits would be rescinded in the future, producing a ripple effect beyond the coal industry. EPA argues that the veto, while highly unusual, is justified because the project involves unacceptable environmental impacts. The agency says that it is not currently reviewing any other previously authorized Appalachian surface coal mining project pursuant to Section 404(c).

The owner of the site, the Mingo Logan Coal Company, challenged EPA's action in federal court, even before the veto was finalized. In March 2012, a federal district court agreed with the industry petitioners and concluded that the CWA does not give EPA the power to render a permit invalid once it has been issued by the Corps. Although the language of 404(c) is "awkwardly written and extremely unclear," the court found EPA's view that it has such authority an unreasonable interpretation of the statute. Thus, it overturned the veto. But in April 2013, a federal appeals court disagreed and reversed the district court's ruling, thus upholding EPA's authority to retroactively veto Section 404 permits.[16] The mining company has petitioned the Supreme Court to review and overturn the appeals court's ruling.

EPA's veto of the Section 404 permit for the West Virginia mine also has drawn congressional attention and criticism. In the 111th and 112th Congresses, legislation was introduced to delete Section 404(c) from the CWA, thus removing EPA's authority to veto permits for projects. In addition, other bills were introduced that were intended to address the veto of the Spruce No. 1 mine project, including proposals to bar EPA from using the 404(c) authority "after the fact," that is, after the Corps has issued a 404 permit; set deadlines for EPA's 404(c) authority; and clarified procedures for elevating 404 permitting decisions to higher level agency and department officials.

Following the court of appeals ruling in April that upheld EPA's authority to retroactively veto a 404 permit, similar bills have been introduced in the 113th Congress to prohibit EPA from using the 404(c) authority after the Corps has issued a permit (H.R. 524 and S. 830).

Nationwide Permits

Nationwide permits are a key means by which the Corps minimizes the burden of its regulatory program. A nationwide permit is a form of general permit which authorizes a category of activities throughout the nation and is valid only if the conditions applicable to the permit are met. These general permits authorize activities that are similar in nature and are judged to cause only minimal adverse effect on the environment, individually and cumulatively. General permits authorize landowners to proceed without having to obtain individual permits in advance.[17]

The current program has few strong supporters, for differing reasons. Developers say that it is too complex and burdened with arbitrary restrictions. Environmentalists say that it does not adequately protect aquatic resources. At issue is whether the program has become so complex and expansive that it cannot either protect aquatic resources or provide for a fair regulatory system, which are its dual objectives.

Nationwide permits are issued for periods of no longer than five years and thereafter must be reissued by the Corps. Most recently, the Corps reissued the nationwide permits in February 2012. This action included reissuing 48 of the 49 existing NWPs, two new NWPs, three new general conditions, and three new definitions, and some modifications of the existing permits, general conditions, and definitions.[18]

The major focus of the reissued permits was what to do with NWP 21, the nationwide permit authorizing certain discharges associated with surface coal mining activities, which has a long and controversial history. In reissuing the final permit in 2012, the Corps added a 1/2-acre and 300- linear foot limit for the loss of stream beds when NWP 21 is used. It strictly prohibits use of NWP 21 to authorize discharges of dredged or fill material into U.S. waters to construct valley fills associated with surface coal mining. With these new limitations, the Corps concluded that the permit will ensure that activities covered by the permit will not result in more than minimal adverse effects on the environment.[19]

Section 404 Judicial Proceedings: SWANCC and Rapanos

The Section 404 program has been the focus of numerous lawsuits, most of which have sought to narrow the geographic scope of the regulatory program.

SWANCC

An issue of long-standing controversy is whether isolated waters are properly within the jurisdiction of Section 404. Isolated waters (those that lack a permanent surface outlet to downstream waters) which are not physically adjacent to navigable surface waters often appear to provide few of the values for which wetlands are protected, even if they meet the technical definition of a wetland. In January 2001, the Supreme Court ruled on the question of whether the CWA provides the Corps and EPA with authority over isolated waters and wetlands. The Court's 5-4 ruling in *Solid Waste Agency of Northern Cook County (SWANCC) v. U.S. Army Corps of Engineers* (531 U.S. 159 (2001)) held that the denial of a Section 404 permit for disposal on isolated wetlands solely on the basis that migratory birds use the site exceeds the authority provided in the CWA. The full extent of retraction of the regulatory program resulting from this decision remains unclear, even more than nine years after the ruling. Environmentalists believe that the Court misinterpreted congressional intent on the matter, while industry and landowner groups welcomed the ruling.[20]

Policy implications of how much the decision restricts federal regulation depend on how broadly or narrowly the opinion is applied, and since the 2001 Court decision, other federal courts have issued a number of rulings that have reached varying conclusions. Some federal courts have interpreted *SWANCC* narrowly, thus limiting its effect on current permit rules, while a few read the decision more broadly. Attorneys for industry and developers say that the courts will remain the primary battleground for CWA jurisdiction questions, so long as neither the Administration nor Congress takes steps to define jurisdiction.

The government's view on the key question of the scope of CWA jurisdiction in light of *SWANCC* and other court rulings came in a legal memorandum issued jointly by EPA and the Corps in January 2003.[21] It provided a legal interpretation essentially based on a narrow reading of the Court's decision, thus allowing federal regulation of some isolated waters to continue (in cases where factors other than the presence of migratory birds may exist, thus allowing for assertion of federal jurisdiction), but it called for more review by higher levels in the agencies in such cases. Administration press releases said that the guidance demonstrates the government's commitment to "no-net-loss" wetlands policy. However, it was apparent that the issues remained under discussion, because at the same time, the Administration issued an advance notice of proposed rulemaking (ANPRM) seeking comment on how to define waters that are under the regulatory program's jurisdiction. The ANPRM did not actually propose rule changes, but it indicated possible ways that Clean Water Act rules might be modified to further limit federal jurisdiction, building on *SWANCC* and some of the subsequent legal decisions. The government received more than 133,000 comments on the ANPRM, most of them negative, according to EPA and the Corps. Environmentalists and many states opposed changing any rules, saying that the law and previous court rulings call for the broadest possible interpretation of the Clean Water Act (and narrow interpretation of *SWANCC*), but developers sought changes to clarify interpretation of the *SWANCC* ruling.

In December 2003, EPA and the Corps announced that the Administration would not pursue rule changes concerning federal regulatory jurisdiction over isolated wetlands. The EPA Administrator said that the Administration wanted to avoid a contentious and lengthy rulemaking debate over the issue. Nonetheless, interest groups on all sides have been critical of confusion in implementing the 2003 guidance, which constitutes the main tool for interpreting the reach of the *SWANCC* decision. Environmentalists remain concerned about diminished protection resulting from the guidance, while developers said that without a new rule, confusing and contradictory interpretations of wetland rules likely will continue. In that vein, a Government Accountability Office (GAO) report concluded that Corps districts differ in how they interpret and apply federal rules when determining which waters and wetlands are subject to federal jurisdiction, documenting enough differences that the Corps undertook a comprehensive survey of its district office practices to help promote greater consistency.[22] Concerns over inconsistent or confusing regulation of wetlands have also drawn congressional interest.[23]

Rapanos-Carabell

Federal courts continue to have a key role in interpreting and clarifying the *SWANCC* decision. In February 2006, the Supreme Court heard arguments in two cases brought by landowners (*Rapanos v. United States*; *Carabell v. U.S. Army Corps of Engineers*) seeking to narrow the scope of the CWA permit program as it applies to development of wetlands. The issue in both cases had to do with the reach of the CWA to cover "waters" that were not navigable waters, in the traditional sense, but were connected somehow to navigable waters or "adjacent" to those waters. (The act requires a federal permit to discharge dredged or fill materials into "navigable waters.") Many legal and other observers hoped that the Court's ruling in these cases would bring greater clarity about the scope of federal regulatory jurisdiction.

The Court's ruling was issued on June 19, 2006 (*Rapanos et al., v. United States*, 547 U.S. 715 (2006)). In a 5-4 decision, a plurality of the Court, led by Justice Scalia, held that the lower court had applied an incorrect standard to determine whether the wetlands at issue are covered by the CWA. Justice Kennedy joined this plurality to vacate the lower court decisions and remand the cases for further consideration, but he took different positions on most of the substantive issues raised by the cases, as did four other dissenting justices.[24] Legal observers suggested that the implications of the ruling (both short-term and long-term) are far from clear. Because the several opinions written by the justices did not draw a clear line regarding what wetlands and other waters are subject to federal jurisdiction, one result has been more case-by-case determinations and continuing litigation. The Senate Environment and Public Works Committee held a hearing on issues raised by the Court's ruling in August 2006. Members and a number of witnesses urged EPA and the Corps to issue new guidance to clarify the scope of the ruling.

On June 5, 2007—nearly one year after the *Rapanos* ruling—EPA and the Corps did issue guidance to enable their field staffs to make CWA jurisdictional determinations in light of the decision. According to the nonbinding guidance, the agencies would assert regulatory jurisdiction over certain waters, such as traditional navigable waters and adjacent wetlands. Jurisdiction over others, such as non-navigable tributaries that do not typically flow year-round and wetlands adjacent to such tributaries, would be determined on a case-by-case basis, to determine if the waters in question have a significant nexus with a traditional navigable water. The guidance details how the agencies should evaluate whether there is a significant nexus. The guidance was not intended to increase or decrease CWA jurisdiction, and it did not supersede or nullify the January 2003 guidance memorandum, discussed above, which addressed jurisdiction over isolated wetlands in light of *SWANCC*.

In accompanying documents, the agencies said that the Administration was considering a rulemaking in response to the *Rapanos* decision, but they noted that developing new rules would take more time than issuing the guidance. They also noted that, while the guidance provides more clarity for how jurisdictional determinations will be made concerning non-navigable tributaries and their adjacent wetlands, legal challenges to the scope of CWA jurisdiction are likely to continue. The guidance was effective immediately, but the agencies also solicited public comments and said that further guidance could be issued in the future. Thus, in December 2008, the Corps and EPA issued revised guidance in an effort to clarify the scope of CWA protection, providing more detail on several issues, including how to identify traditional navigable waters and adjacent wetlands. The guidance took the view that waters are jurisdictional if they satisfy *either* the plurality or Kennedy tests in *Rapanos*. The 2008 guidance also updated the 2007 guidance with more detail for determining whether a wetland is adjacent to a traditional navigable water and whether a tributary of a navigable water is subject to the act—key issues raised by the *Rapanos* decision.

In April 2011, the Obama Administration weighed into the CWA jurisdiction debate as EPA and the Corps proposed new joint agency guidance to clarify regulatory jurisdiction over U.S. waters and wetlands and to replace the agencies' 2008 guidance. Like the existing guidance, the proposed revisions would adopt the Kennedy-test-or-plurality-test view of interpreting Rapanos. However, the agencies believed that a wider evaluation of jurisdiction is possible than the existing guidance suggests, stating, "after careful review of these opinions, the agencies concluded that previous guidance did not make full use of the authority provided by the CWA to include waters within the scope of the Act, as interpreted by the Court."[25]

EPA and the Corps acknowledged that, compared with the existing guidance, the proposed revisions were likely to increase the number of waters identified as protected by the CWA—a conclusion that pleased some observers but alarmed others. Supporters believed that new guidance will improve protection of U.S. waters and wetlands, while critics argue that it represents over-reaching, beyond authority provided by Congress. Others faulted the continued reliance on federal guidance, which is not binding and lacks the force of law, yet can have significant impact on regulated entities. Final guidance was submitted to the White House Office of Management and Budget (OMB) for review in February 2012, but it was not released. Instead, in September 2013, EPA and the Corps withdrew the revised guidance document from OMB and submitted draft regulations to OMB for interagency review. The substance of this proposal, and when it might be proposed, are unknown for now.

SWANCC and *Rapanos* generated confusion beyond what already existed as to the reach of "waters of the United States." The lack of a majority rationale in *Rapanos* has led lower courts to extract different tests from the decision for measuring this reach, and Justice Kennedy's "significant nexus" concept remains amorphous and undefined. The EPA-Corps 2011 draft guidance—now withdrawn—was intended to reduce the confusion, but many observers and stakeholders contend that jurisdictional issues remain in dispute throughout the country, leading to costly project delays and uncertain protection of wetland resources.

While the issue of how regulatory protection of wetlands is affected by the *SWANCC* and *Rapanos* decisions continues to evolve, the remaining responsibility to protect affected wetlands falls on states and localities. Whether states will act to fill in the gap left by removal of some federal jurisdiction is likely to be constrained by budgetary and political pressures, but after *SWANCC* a few states (Wisconsin and Ohio, for example) passed new laws or amended regulations to do so. In comments on the 2003 ANPRM, many states said that they do not have authority or financial resources to protect their wetlands, in the absence of federal involvement.

Congressional Response

Legislation to reverse the *SWANCC* and *Rapanos* decisions was introduced on several occasions since the 107th Congress. In the 111th Congress, the Senate Environment and Public Works Committee approved S. 787—the first such proposal to advance from a congressional committee, but no further legislative action occurred.

As approved by the Senate committee, the bill would have amended the CWA to define "waters of the United States" and to use this term to define the jurisdictional reach of the act. The term would have been defined to mean:

> all waters subject to the ebb and flow of the tide, the territorial seas, and all interstate and intrastate waters, including lakes, rivers, streams (including intermittent streams), mudflats, sandflats, wetlands, sloughs, prairie potholes, wet meadows, playa lakes, and natural ponds, all tributaries of any of the above waters, and all impoundments of the foregoing.

The bill would have excluded prior converted cropland and certain waste treatment systems from the term "waters of the United States" (terms now covered by regulatory exemptions), and it would have protected, or saved, existing statutory exclusions such as for dredge or fill discharges from normal farming activities. During markup, the committee rejected several amendments that would have struck some of the terms in the new definition (such as mudflats and prairie potholes), but it approved language stating that the

CWA's jurisdiction shall be construed consistent with EPA and Corps interpretation as of January 8, 2001, the day before the *SWANCC* ruling, and consistent with Congress' constitutional authority.

Proponents of the Senate committee legislation have argued that Congress must clarify the important issues left unsettled by the Supreme Court's 2001 and 2006 rulings and by the Corps/EPA guidance. Bill supporters argued that the legislation would "reaffirm" what Congress intended when the CWA was enacted in 1972 and what EPA and the Corps have subsequently been practicing until recently, in terms of CWA jurisdiction. It also would have deleted the word "navigable" from the act, replaced by the term "waters of the United States," in order to clarify that Congress intends the purpose of the law as broadly protecting the quality of the nation's waters, not just sustain navigability in the traditional sense. But critics asserted that the legislation would expand federal authority, and thus would have unintended but foreseeable consequences that are likely to increase confusion, rather than settle it. Critics questioned the constitutionality of the bill, arguing that, by broadly including U.S. waters in the jurisdiction of the CWA, it would exceed the limits of Congress's authority under the Constitution. The version approved by the Senate committee included language stating that the bill shall be construed consistently with "the legislative authority of Congress under the Constitution."

Companion House legislation was introduced in the 111th Congress (H.R. 5088).[26] Like S. 787, the House bill was intended to clarify regulatory scope of the CWA and restore jurisdiction as it had been interpreted prior to the *SWANCC* and *Rapanos* rulings. Like the Senate committee bill, H.R. 5088 would have deleted the word "navigable" from the law and would have amended the CWA to define "waters of the United States," which would become the operational term for jurisdiction. Unlike the Senate committee bill described above, the new definition of that term would have been drawn from existing EPA-Corps regulatory definitions, with some modifications. The bill was criticized based on concern that it would increase the scope of federal jurisdiction, not merely re-state what Congress enacted in 1972. There was no further legislative action on the bill.

Officials of the Obama Administration are on record as favoring legislation that would clarify waters protected by the CWA, but they did not develop or endorse any specific proposal. In May 2009, Administration officials sent letters to House and Senate committee leaders outlining principles for such legislation and urging Congress to consider four general principles:

- Broadly protect the nation's waters;
- Make the definition of covered waters predictable and manageable;
- Promote consistency between CWA and agricultural wetlands programs; and
- Recognize long-standing practices, such as exemptions now in effect through regulations and guidance.[27]

In light of the widely differing views of proponents and opponents, future prospects for legislation on the geographic scope of CWA jurisdiction are highly uncertain. Critics have questioned the constitutionality of legislation that has been proposed, and have asserted that it would expand federal authority, thus likely increasing confusion, rather than settling it. An additional difficulty of legislating changes to the CWA in order to protect wetlands results from the fact that the complex scientific questions about such areas (see discussion above, "Wetlands: Science and Information") are not easily amenable to precise resolution in law. The debate over revising the act highlights the challenges of using the law to try to do so.

Efforts by EPA and the Corps to develop revised *Rapanos* guidance (which has now been withdrawn) have been controversial in Congress and elsewhere. Legislative provisions to prohibit the agencies from funding activities related to revising the guidance were included in several appropriations bills in the 112th Congress, but none of these provisions was enacted. Interest in similar legislation concerning the guidance has continued with bills in the 113th Congress, such as S. 1006 and H.R. 1829, to prevent the agencies from finalizing the 2011 draft guidance, and S. 890/H.R. 3377, which would amend the CWA with a narrow definition of waters that are subject to the act's jurisdiction.

Should All Wetlands Be Treated Equally?

Under the Section 404 program, there is a perception that all jurisdictional wetlands are treated equally, regardless of size, functions, or values. In reality, this is not the case, because the Corps' general permits do provide accelerated regulatory decisions for many activities that affect wetlands. Further, a number of types of activities are fully exempt from 404 permit requirements as a result of statutory provisions enacted in 1977 (including ongoing farming, ranching, and forestry activities, as specified in Section 404(f)), and regulatory exemptions (including for prior converted croplands, which are wetlands that were drained, dredged, filled, leveled, or otherwise manipulated before

December 23, 1985, to make production of an agricultural commodity possible).

However, this perception has led critics to focus on situations where a wetland has little apparent value, but the landowner's proposal is not approved, or the landowner is penalized for altering a wetland without a federal permit. Critics believe that one possible solution may be to have a tiered approach for regulating wetlands. Legislation introduced in past Congresses proposed to establish multiple tiers (typically three)—from highly valuable wetlands that should receive the greatest protection to the least valuable wetlands where alterations might usually be allowed. Some states (New York, for example) use such an approach for state-regulated wetlands.

Three questions arise: (1) What are the implications of implementing a classification program? (2) How clearly can a line separating each wetland category be defined? (3) Are there regions where wetlands should be treated differently? Regarding classification, even many wetland protection advocates acknowledge that there are some situations where a wetland designation with total protection is not appropriate. But they fear that classification for different degrees of protection could be a first step toward a major erosion in overall wetland protection. Also, these advocates would probably like to see almost all wetlands presumed to be in the highest protection category unless experts can prove an area should receive a lesser level of protection, while critics who view protection efforts as excessive would seek the reverse. In response to these concerns, Corps and EPA officials note that existing guidance and regulations already provide substantial flexibility to implement current programs, allowing, for example, less vigorous permit review to small projects with minor environmental impacts. Some types of wetlands are already treated differently—for example, playas and prairie potholes, which have somewhat different definitions under swampbuster (discussed below). However, this differential treatment contributes to questions about federal regulatory consistency on private property.

Locating the boundary line of a wetland can be controversial when the line encompasses areas that do not meet the image held by many. Controversy would likely grow if a tiered approach required that lines segment wetland areas. On the other hand, a consistent application of an agreed-on definition might lead to fewer disputes and result in more timely decisions.

Some states have far more wetlands than others. Different treatment has been proposed for Alaska, because about one-third of the state is designated as wetlands, yet a very small portion has been converted. In the past, legislative proposals have been made to exempt that state from the Section 404 program until 1% of its wetlands have been lost.

Agriculture and Wetlands

National surveys more than two decades ago indicated that agricultural activities had been responsible for about 80% of wetland loss in the preceding decades, making this topic a focus for policymakers seeking to protect the remaining wetlands. Congress responded by creating wetland conservation programs in farm legislation starting in 1985.

Conservation programs in the farm bill use both disincentives and incentives to encourage landowners to protect and restore wetlands. Swampbuster and the Wetlands Reserve Program are the two largest efforts, but others such as the Conservation Reserve Program's wetland and buffer acres pilot program and the Conservation Reserve Enhancement Program are also being used to protect wetlands. The 110th Congress reauthorized farm programs through FY2012 (Food, Conservation, and Energy Act of 2008, P.L. 110-246). Wetlands also were a major topic of discussion in debate on this bill, which authorized new programs that could further assist wetlands conservation.

Members of the farm community have expressed a wide range of views about wetland protection, from strong opposition to strong support. These views are frequently framed in the context of two general concerns about wetland protection efforts. First, as a philosophical matter, some object to federal regulation of private lands, regardless of the societal values those lands might provide. Second, many farmers want certainty and predictability about the land they farm to limit their financial risk. Therefore, if wetlands are located on farm property, they want assurances that the boundary line delineating wetlands will remain where located for as long as possible.

Swampbuster

Swampbuster, enacted in 1985, uses disincentives rather than regulations to protect wetlands on agricultural lands. It removes a farmer's eligibility from all government price and income support programs for activities such as draining, dredging, filling, leveling or otherwise altering a wetland. Swampbuster has been controversial with farmers concerned about redefining an appropriate federal role in wetland protection on agricultural lands, and with wetland protection advocates concerned about inadequate enforcement.

Since 1995, the NRCS has made wetland determinations only in response to requests because of uncertainty over whether changes in regulation or law would modify boundaries that have already been delineated. NRCS has estimated that more than 2.6 million wetland determinations have been made and that more than 4 million may eventually be required.

Swampbuster amendments in 1996 (P.L. 104-127) granted producers greater flexibility by making changes, such as: exempting swampbuster penalties when wetlands are voluntarily restored; providing that prior converted wetlands are not to be considered "abandoned" if they remain in agricultural use; and granting good-faith exemptions. They also encourage mitigation, established a mitigation banking pilot program, and repealed required consultation with the FWS. Amendments enacted in the 2008 farm bill require an additional layer of review within USDA for compliance with swampbuster.

The 113th Congress is considering legislation to renew the farm bill, including limited modifications to the swampbuster program. In June 2013, the Senate passed S. 954 with a provision that would add the federally funded portion of crop insurance premiums to the list of program benefits that could be lost if a producer is found to convert a wetland to crop production. The House-passed bill (H.R. 2642) includes no comparable provision.[28]

Other Agricultural Wetlands Programs

Several USDA conservation programs provide federal payments to private agricultural landowners for changes in land use or management to achieve environmental benefits, including wetlands protection.[29] Under the Wetland Reserve Program (WRP), enacted in 1990, landowners receive payments for placing easements on farmed wetlands. It provides long-term technical and financial assistance to landowners with the opportunity to protect, restore, and enhance wetlands on their property, and to establish wildlife practices and protection. WRP offers permanent easements that pay 100% of the value of an easement and up to 100% of easement restoration costs, and 30-year easements that pay up to 75% of the value of an easement and up to 75% of easement restoration costs. WRP also offers restoration cost-share agreements to restore wetland functions and values without placing an easement on enrolled acres. Through FY2012, projects totaling nearly 2.6 million acres have been enrolled in the program. Most of the land is enrolled under permanent or 30-year easements, while only about 10% is enrolled under 10-year restoration cost-share agreements, according to NRCS.

Strong farmer interest led Congress to raise the WRP enrollment ceiling in both the 2002 and 2008 farm bills. The 2008 legislation increased the WRP maximum enrollment cap from 2.275 million acres to 3.014 million acres and expanded eligible lands to include certain types of private and tribal wetlands, croplands, and grasslands, as well as lands that meet the habitat needs of wildlife species. The bill made certain program changes, including specifying criteria for ranking program applications, and requiring USDA to submit a report to Congress on long-term conservation easements under the program. The legislation authorized a new Wetlands Reserve Enhancement Program, allowing USDA to enter into agreements with states in order to leverage federal funds for wetlands protection and enhancement.[30]

The 2002 farm bill expanded the 500,000-acre wetland and buffer acreage pilot program within the Conservation Reserve Program (CRP) to a 1-million-acre program available nationwide. CRP allows producers to enter into 10- to 15-year contracts to install certain conservation practices. The 2008 farm bill amended the pilot program to increase the amount of acreage that states can enroll (up to 100,000 acres, or a national maximum of 1 million acres). Participants must agree to restore wetland hydrology, establish appropriate vegetation, and refrain from commercial use of the land. The wetland and buffer program may be an important to overall protection efforts in the wake of the *SWANCC* decision, discussed above, which limited the reach of the Section 404 permit program to many small wetlands that are isolated from navigable waterways. As of October 2012, 27 million acres were enrolled in this program through more than 698,000 contracts on approximately 390,000 farms.

In 2004, USDA announced a Non-Floodplain Wetland Restoration Initiative to allow enrollment of up to 250,000 acres of large wetland complexes and playa lakes located outside the 100-year floodplain in the CRP. As of April 2012, there were 154,000 acres enrolled. USDA also established a Floodplain Wetland Restoration Initiative to enroll wetlands located in the 100-year floodplain in the CRP. As of April 2012, a total of 213,000 acres were enrolled. Participants receive incentive payments equal to 25% of the cost to help pay for restoring the hydrology of the site, as well as rental payments and cost sharing assistance to install eligible conservation practices.

The 2008 farm bill included amendments affecting several agriculture conservation programs, including the Environmental Quality Incentives Program (EQIP), the Farmland Protection Program (FPP), and the Wildlife Habitat Incentive Program (WHIP), in ways that may have incidental protection benefits for wetlands, because of higher funding levels or because of program changes. For example, EQIP supports the installation or implementation of structural and management practices, and the 2008 farm bill expanded the program to include practices that enhance wetlands. Finally, some programs could less directly help protect wetlands, including the Conservation Security Program (renamed the Conservation Stewardship Program), which provides payments to install and maintain practices on agricultural lands; the new Agricultural Water Enhancement Program (replacing the previous Ground and Surface Water Conservation Program; it is funded through EQIP), which is designed to address water quality and quantity concerns on agricultural land; and several other programs to better manage water resources.[31]

Farm bill legislation in the 113th Congress would modify these programs in several respects. Under both S. 954, passed by the Senate in June, and H.R. 2642, passed by the House in July, the WRP and FPP would be repealed and consolidated in a new conservation program, the Agricultural Conservation Easement Program (ACEP). The new program would retain most of the program provisions in the current WRP by establishing an easement program to protect and restore wetlands. Both bills would reauthorize EQIP with a 5% funding carve-out for wildlife habitat practices (similar to WHIP, which would be repealed).[32]

Agricultural Wetlands and the Section 404 Program

The CWA Section 404 program applies to qualified wetlands in all locations, including agricultural lands. But the Corps and EPA exempt "prior converted lands" (wetlands modified for agricultural purposes before 1985) from Section 404 permit requirements under a memorandum of agreement (MOA), and since 1977 the Clean Water Act has exempted "normal farming activities." The Supreme Court's *SWANCC* decision exempts certain isolated wetlands from Corps jurisdiction; NRCS estimated that about 8 million acres in agricultural locations might be exempted by this decision.

While these exemptions and the MOA displease some protection advocates, they probably dampened some of the criticism from farming interests over federal regulation of private lands. On the other hand, the prospect that Congress might enact legislation to reverse the Court's 2001 and 2006 rulings, discussed above, has particularly alarmed farm groups, who fear that changes in law or regulations could negatively affect their activities. Because of differences between the CWA and farm bill on the jurisdictional status of certain wetlands (e.g., isolated wetlands may be regulated differently by federal agencies), in 2005 the Corps and NRCS signed a Memorandum of

Understanding and issued joint guidance clarifying circumstances where wetlands delineation made by one agency can be accepted for determining the jurisdiction of the other agency.[33] Some of the wetlands that fall outside Section 404 requirements as a result of judicial decisions can now be protected if landowners decide to enroll them into the revised farmable wetlands program or under other new initiatives.

Other Federal Protection Efforts

Many federal agencies have been active in wetland improvement efforts in recent years. In particular, the Fish and Wildlife Service (FWS) has been promoting the success of its Partners for Fish and Wildlife program, which Congress reauthorized through FY2011 in 2006 (P.L. 109-294). Through voluntary agreements, the Partners program provides technical assistance and cost share incentives directly to landowners for wetland restoration projects on private lands.[34]

FWS also administers the National Coastal Wetlands Conservation Grant Program. Under this program, federal grants, matched by state and local contributions, as well as from private landowners and conservation groups, are used to acquire, restore, or enhance coastal wetlands and adjacent uplands to provide long-term conservation benefits to fish, wildlife, and their habitats.

The federal government generally provides 50% of the total costs of a project, but the federal share can be increased to 75% if the state maintains a fund for acquiring coastal wetlands. Since 1992, about $183 million in grants have been awarded to 25 coastal states and one U.S. territory for projects involving 250,000 acres of coastal wetland ecosystems.[35]

Other programs also restore and protect domestic and international wetlands. One of these derives from the North American Wetlands Conservation Act, reauthorized through FY2012 in P.L. 109-322 with an appropriations ceiling of $75 million annually. This act provides grants for wetland conservation projects in Canada, Mexico, and the United States. The FWS has combined funding for this program with several other laws into what it calls the North American Wetlands Conservation Fund. According to the FWS, since the program began in 1991, the United States and its 4,800 domestic and international partners have conserved, restored, or enhanced 27 million acres of wetlands in the three countries, equivalent to an area larger than the state of Tennessee.

Under the Convention on Wetlands of International Importance, more commonly known as the Ramsar Convention, the United States is one of 168 nations that have agreed to slow the rate of wetlands loss by designating wetland sites of international importance. These nations have designated 2,169 sites, totaling 509 million acres, since the convention was adopted in 1971. The United States has designated 35 sites pursuant to the convention, encompassing 4.5 million acres.

Private Property Rights and Landowner Compensation

An estimated 74% of all remaining wetlands in the conterminous states are on private lands. Questions of federal regulation of private property stem from the argument that land owners should be compensated when a "taking" occurs and alternative uses are prohibited or restrictions on use are imposed to protect wetland values. The U.S. Constitution provides that property owners shall be compensated if private property is "taken" by government action. The courts generally have found that compensation is not required unless all reasonable uses are precluded. Many individuals or companies purchase land with the expectation that they can alter it. If that ability is denied, they contend, then the land is greatly reduced in value. Many argue that a taking should be recognized when a site is designated as a wetland. In 2002, the Supreme Court held that a Rhode Island man, who had acquired property after the state enacted wetlands regulation affecting the parcel, is not automatically prevented from bringing an action to recover compensation from the state. Instead, the court ruled that the property retained some economic use after the state's action. *(Palazzolo v. Rhode Island*, 533 U.S. 606 (2002)).

Congress has explored these wetlands property rights issues on several occasions. An example is an October 2001 hearing by the House Transportation and Infrastructure Committee, Subcommittee on Water Resources and the Environment.[36] Recent Congresses considered, but did not enact, property rights protection proposals.

STATE PROTECTION EFFORTS

In addition to federal programs and activities, wetlands in the United States are regulated and protected through a variety of state and local laws and regulations, as well as through initiatives and actions of nongovernmental organizations, schools and universities, and private citizens. The role of states in wetland protection is especially important, as noted in a study by the Environmental Law Institute.

> States have long held the right and the responsibility to provide stewardship over their resources, and state agency staff typically have a well-versed understanding of the "lay of the land," in terms of both topography and state priorities, policies, and practices. Finally, in light of recent uncertainty over federal jurisdiction of wetlands and limited federal resources for wetland protection, the role of states in conserving wetlands may be more important now than ever before.[37]

States use a variety of programs and tools to protect and manage wetlands, including regulation and mitigation, wetland water quality standards, monitoring and assessment, voluntary restoration, tax incentives, coordination among state and federal agencies, and public/private partnerships. Programs vary substantially from state to state and often derive their authorities from more than one statute and/or regulation. As a result, different programs may be administered by different state agencies. In addition, programs may change from year to year.[38]

Every state regulates, to some degree, activities that affect wetlands, but two-thirds of the states lack regulatory programs that *comprehensively* regulate wetlands. Many states rely solely or primarily on authority in CWA Section 401, under which states may review any activity that requires a federal permit or license to determine its effect on the state's water quality standards.[39] Section 401 gives states the authority to approve, condition, or deny the federal permit— including a Section 404 permit—or license based on their review. In areas where there is no Section 404 permit requirement, and therefore no opportunity for review under Section 401, some states also require a state permit for activities that affect aquatic resources: 23 states have authority to issue permits for dredge and fill activities in wetlands and other waters of the state, such as geographically isolated wetlands (although as described previously, only New Jersey and Michigan have been delegated 404 permitting authority).

As is the case with the federal regulatory program under CWA Section 404, an important consideration is how a state determines which waters fall within its regulatory jurisdiction. States' definitions of their waters are typically much broader than the federal definition of "waters of the United States," meaning that states may exert jurisdiction over waters within their boundaries that are not covered by the CWA. State definitions often includes phrases such as "all surface waters." They also may exclude certain waters, such as private lakes or ponds. Groundwater is not included in the federal regulatory definition, but most states include groundwater in their regulatory programs.[40] All 50 states include wetlands in either or both their statutory and regulatory definitions of state waters—32 make this inclusion explicit, and 18 define waters more generally, including wetlands implicitly. The inclusion of wetlands in a state's definition of state waters does not give automatic protection to these waters; the state must also have some form of complementary regulatory authority, such as to issue permits.[41]

Other findings of the ELI report include the following.

- The majority of states have adopted legislation, policies, and/or guidelines for mitigating impacts to aquatic resources that are permitted in their states. Mitigation provisions range from general requirements to specific replacement ratios, site preferences, and mitigation options such as purchasing credits from a mitigation bank (also see "Wetland Restoration and Mitigation").
- One-third of states report having a wetland-specific monitoring and/or assessment program or monitoring wetlands as part of a larger state monitoring program.
- Nearly one-half of the states operate a formal program for partnering with private landowners on restoration or conservation, and a majority of states report that they conduct outreach or provide technical assistance to private landowners. Ninety percent of states have one or more agencies that carry out education and outreach activities related to wetlands.

THE LOUISIANA EXPERIENCE

Much of the attention to reverse wetland loss has focused on Louisiana, where an estimated 80% of the total loss of U.S. coastal wetlands has occurred and where about 40% of U.S. coastal wetlands remaining in the lower 48 states are located (coastal wetlands are about 5% of all U.S. wetlands). Changes to Louisiana's coastal area result from a combination of natural environmental processes (erosion, saltwater intrusion into fresh systems, sea level rise) and human-related activities, according to the U.S. Geological Survey (USGS). Wetland loss has occurred naturally for centuries, but until recently, land losses have been counterbalanced by various natural wetland-building processes.

USGS estimates that, since 1932, coastal Louisiana has experienced a net change in land area of approximately 1,883 square miles—an area the size of Delaware. Land loss rates on the Louisiana coast have slowed from an average of more than 30 square miles per year between 1956 and 1978, to an estimated 11.8 square miles per year from 1985 to 2004. When the hurricanes of 2005 and 2008 are factored in, the trend increased the amount of land loss to 16.6 square miles from 1985 to 2010. According to USGS, if this loss were to occur at a constant rate, it would equal losing more than a football field every hour.[42] As a result of wetlands loss, the natural flow of Mississippi River and floodwaters to feed sediment to the marshes has been reduced. Saltwater has invaded the brackish estuaries, destroying vegetation and areas that are needed for fish, shellfish, and wildlife. In response to these losses, Congress authorized a task force, led by the Corps, to prepare a list of coastal wetland restoration projects in the state, and also provided funding to plan and carry out restoration projects in this and other coastal states under the Coastal Wetlands Planning, Protection and Restoration Act of 1990, also known as the Breaux Act.[43] The projects range from reintroduction of freshwater and diversion of sediment to construction of shoreline barriers and planting of vegetation. In total, the estimated total cost to complete all 147 approved projects is $1.78 billion.

In a 2007 report, GAO reported that it is impossible to determine the collective success of restoring coastal wetlands in Louisiana, because of an inadequate approach to monitoring. GAO had reviewed the Breaux Act program to identify the types of projects that have been designed and lessons that have been learned from 74 projects that have been completed so far.[44] Others, including the National Oceanic and Atmospheric Administration, disagreed with GAO's findings, observing that long-term data being provided through ongoing project monitoring are intended to yield insight into qualitative and quantitative project performance.

In the wake of hurricanes Katrina and Rita in 2005, multiple legislative proposals were introduced to fund additional restoration projects already planned by the U.S. Army Corps of Engineers and to explore other opportunities that would restore and stabilize wetlands in southern Louisiana. Before the hurricanes, Congress was considering legislation that would have provided about $2 billion to the restoration effort. Since the 2005 hurricanes, more expansive options costing up to $14 billion that were proposed in the 1998 report *Coast 2050* have also been considered.[45] The Gulf of Mexico

Energy Security Act, legislation that authorizes additional revenues to states adjacent to offshore oil and gas production activities, was passed during the final days of the 109th Congress.[46] One of the purposes for which these revenues can be spent is wetland restoration, and the availability of these funds may affect the amount and scale of wetland restoration activity in the central Gulf Coast.

Concern for Louisiana's coastal wetlands was heightened by the oil spill following the April 2010 explosion of BP's drilling rig, the Deepwater Horizon, in the Gulf of Mexico. Although efforts focused on preventing oil from reaching coastal shorelines, some oil escaped capture and was pushed by wind and tides towards land. The degrees of impacts of oil on wetland vegetation are variable and complex and can be both acute and chronic, ranging from short-term disruption of plant functioning to mortality. The primary acute damage to the marshes is that plants, which hold the soil in place and stabilize shoreline, suffocate and die, especially if multiple coatings of oil occur. Once vegetation dies, the soil collapses. Then the soil becomes flooded, and plants cannot re-grow. If plants cannot re-establish, soil erosion is accelerated, giving rise to even more flooding and further wetland loss. If oil penetrates into the sediments, roots are continuously exposed to oil, with chronic toxicity making production of new shoots problematic.

Consequently, plant recovery is diminished, and eventually land loss occurs. In addition to direct impacts on plants, oil that reaches wetlands also affects animals that use wetlands during their life cycle, especially benthic organisms which reside in the sediments and are a foundation of the food chain.[47]

Public and private efforts were taken to protect the wetlands from oil that moved through Gulf waters towards coastal areas, but scientists remained concerned that high tides and wind could push oil into the marshes, and that the grasses and other vegetation that provide habitat for fish and wildlife would likely be destroyed. Wetland plants can be affected both by oil that floats over the surface of the marsh and by oil that has been incorporated into sediment. While oil was still flowing from the Deepwater Horizon site, cleanup of marshes was limited to triage of heavily oiled marshes and wetlands, because experts were concerned that greater harm than good could be done to the sensitive environmental ecosystems. The well was capped and oil stopped flowing from the well site in mid-July 2010. Experts say that spill response efforts succeeded in keeping large amounts of oil from reaching coastal marshes. Nevertheless, oil remains in the Gulf environment, and potential for re-oiling of coastal areas, for example as a result of storms, will remain a concern for some time.[48]

A recent federal report observes that Louisiana's *Coast 2050* is a comprehensive plan to protect and restore the state's coastal wetlands, but that other Gulf of Mexico states are only beginning similar planning processes for restoration of the damage caused by the Deepwater Horizon spill.[49]

WETLAND RESTORATION AND MITIGATION

Mitigation has become an important cornerstone of the Section 404 program in recent years. A 1990 MOA signed by the agencies with principal regulatory responsibilities (EPA and the Corps) outlines a sequence of three steps leading to mitigation: first, activities in wetlands should be avoided when possible; second, when they cannot be avoided, impacts should be minimized; and third, where minimum impacts are still unacceptable, mitigation is appropriate. Therefore, mitigation may be required as a condition of a Section 404 permit.

Federal wetland policies during the past 30 years have increasingly emphasized restoration of wetland areas. Much of this restoration occurs as part of efforts to mitigate the loss of wetlands at other sites. The mitigation concept has broad appeal, but implementation has left a conflicting record. Examination of this record, presented in a June 2001 report from the National Research Council, found it to be wanting. The NRC report said that mitigation projects called for in permits affecting wetlands were not meeting the federal government's "no net loss" policy goal for wetlands function.[50] Likewise, a 2001 GAO report criticized the ability of the Corps to track the impact of projects under its current mitigation program that allows in-lieu-fee mitigation projects in exchange for issuing permits allowing wetlands development.[51] Both scientists and policymakers debate whether it is possible to restore or create wetlands with ecological and other functions equivalent to or better than those of natural wetlands that have been lost over time. Results so far seem to vary, depending on the type of wetland and the level of commitment to monitoring and maintenance. Congress has repeatedly endorsed mitigation in recent years.

Some wetland protection advocates are critical of mitigation, which they view as justifying destruction of wetlands. They believe that the Section 404 permit program should be an inducement to avoid damaging wetland areas. These critics also contend that adverse impacts on wetland values are often not fully mitigated and that mitigation measures, even if well-designed, are not adequately monitored or maintained. Supporters of current efforts counter that they generally work as envisioned, but little data exist to support this view. Questions about implementation of the 1990 MOA and controversies over the feasibility of compensating for wetland losses further complicate the wetland protection debate.

In response to criticism in the NRC and GAO reports on mitigation, in 2001, the Corps issued new guidance to strengthen the standards on compensating for wetlands lost to development. But the guidance was criticized by environmental groups and some Members of Congress for weakening rather than strengthening mitigation requirements and for the Corps' failure to consult with other federal agencies. In 2002, the Corps and EPA released an action plan including 17 items that both agencies believed would improve the effectiveness of wetlands restoration efforts.[52]

In 2008, the Corps and EPA promulgated a mitigation rule to replace the 1990 MOA with clearer requirements on what will be considered a successful project to compensate for wetlands lost to activities like construction, mining, and agriculture.[53] The rule sets performance standards and criteria for three types of wetlands mitigation: mitigation banks, in-lieu programs, and permitteeresponsible compensatory mitigation. It sets standards to mitigate the loss of wetlands and associated aquatic resources and is intended to improve the planning, implementation, and management of compensatory mitigation projects designed to restore aquatic resources that are affected by activities that disturb a half-acre or more of wetlands. It also is designed to help ensure no net loss of wetlands by addressing key recommendations raised in the 2001 NRC report. Under the rule, all compensation projects must have mitigation plans that include 12 fundamental components, such as objectives, site selection criteria, a mitigation work plan, and a maintenance plan.[54]

The concept of "mitigation banks," in which wetlands are created, restored, or enhanced in advance to serve as "credits" that may be used or acquired by permit applicants when they are required to mitigate impacts of their activities, is widely endorsed and is the preferred option under the 2008 mitigation rule. Numerous public and private banks have been established, but many believe that it is too early to assess their success. In a study of mitigation, the Environmental Law Institute determined that as of 2005, there were 330 active banks, 75 sold out banks, and 169 banks seeking approval to operate.[55] Provisions in several laws, such as the 1996 farm bill and the 1998 Transportation Equity Act (TEA-21), endorse the mitigation banking concept. In 2003, Congress enacted wetlands mitigation provisions as part of the

FY2004 Department of Defense (DOD) authorization act (P.L. 108-136). Section 314 of that act directed DOD to make payments to wetland mitigation banking programs in instances where military construction projects would result or could result in destruction of or impacts to wetlands.

End Notes

[1] Two places to view material on some of the changes in scientific knowledge and understanding are through the products of the Society of State Wetlands Managers http://www.aswm.org and the Society of Wetland Scientists http://www.sws.org.
[2] Thomas E. Dahl, *Status and Trends of Wetlands in the Conterminus United States, 2004-2009*, U.S. Department of the Interior, Fish and Wildlife Service, 2011, 108p.
[3] For information, see http://www.fws.gov/wetlands
[4] Thomas E. Dahl and Susan-Marie Stedman, *Status and Trends of Wetlands in the Coastal Watersheds of the Conterminous United States 2004-2009*, U.S. Department of the Interior, Fish and Wildlife Service, and National Ocean and Atmospheric Administration, National Marine Fisheries Service, October 2013, 46 p.
[5] See http://water
[6] Office of the President, Council on Environmental Quality, *Conserving America's Wetlands 2008: Four Years of Progress Implementing the President's Goal*, April 2008.
[7] Dahl, T.E. *Status and Trends of Wetlands in the Conterminus United States, 2004-2009*, U.S. Department of the Interior, Fish and Wildlife Service, 2011, p. 86.
[8] Matthew L. Kirwan, Glenn R. Guntenspergen, and Andrea D'Alpaos et al., "Limits on the adaptability of coastal marshes to rising sea level," *Geophysical Research Letters*, vol. 37, no. L23401 (December 2010).
[9] While tidal wetlands do effectively sequester carbon, some wetlands (especially those with low salinity levels) also are a source of GHGs by emitting methane, which is approximately 21 times more powerful as a GHG than CO_2.
[10] Stephen Crooks, Dorothee Herr, and Jerker Tamelander et al., Mitigating Climate Change through Restoration and Management of Coastal Wetlands and Near-shore Marine Ecosystems, Challenges and Opportunities, The World Bank Environment Department, March 2011, http://siteresources.worldbank.org/ENVIRONMENT/Resources/ MtgtnCCthru MgtofCoastalWetlands.pdf.
[11] Stephen Emmett-Mattox, Stephen Crooks, and Jette Findsen, "Wetland Grasses and Gases: Are Tidal Wetlands Ready for the Carbon Markets?," *National Wetlands Newsletter*, November-December 2010, pp. 6-10.
[12] For additional information, see CRS Report RL31411, *Controversies over Redefining "Fill Material" Under the Clean Water Act*, by Claudia Copeland.
[13] The updated list of wetland plants is available at http://wetland_plants.usace.army.mil.
[14] U.S. Environmental Protection Agency, "Final Determination of the Assistant Administrator for Water Pursuant to Section 404(c) of the Clean Water Act Concerning the Spruce No. 1 Mine, Logan County, WV; Notice," 76 *Federal Register* 3126-3128, January 19, 2011. The Final Determination and related materials are available at http://water
[15] There is some ambiguity about whether it is the first EPA veto of a previously issued permit. Initially, EPA said that the agency had never done so, but agency officials subsequently

Wetlands: An Overview of Issues 37

have cited the 1980 veto of a project in North Miami Beach, Florida as having involved a previously authorized project. The facts of that case, concerning a veto of a proposed permit amendment for the project, differ from the Spruce No. 1 mine case.

[16] Mingo Logan Coal Co. v. EPA, 714 F.3d 608 (D.C. Cir. 2013).

[17] For information, see CRS Report 97-223, *The Army Corps of Engineers' Nationwide Permits Program: Issues and Regulatory Developments*, by Claudia Copeland.

[18] For the full text, see http://www.usace.army.mil/Missions/CivilWorks /Regulatory Programand Permits/ NationwidePermits.aspx.

[19] For background, see CRS Report RS21421, *Mountaintop Mining: Background on Current Controversies*, by Claudia Copeland.

[20] For additional information, see CRS Report RL30849, *The Supreme Court Addresses Corps of Engineers Jurisdiction Over "Isolated Waters": The SWANCC Decision*, by Robert Meltz.

[21] See http://www.epa.gov/owow/wetlands

[22] U.S. Government Accountability Office, *Corps of Engineers Needs to Evaluate Its District Office Practices in Determining Jurisdiction*, GAO-04-297, February 2004, 45 pp.

[23] U.S. Congress, House of Representatives, Committee on Transportation and Infrastructure, Subcommittee on Water Resources and Environment, *Inconsistent Regulation of Wetlands and Other Waters*, Hearing 108-58, 108th Cong., 2nd sess., March 30, 2004.

[24] For additional information, see CRS Report RL33263, *The Wetlands Coverage of the Clean Water Act (CWA): Rapanos and Beyond*, by Robert Meltz and Claudia Copeland.

[25] U.S. Environmental Protection Agency and Army Corps of Engineers, "Draft Guidance on Identifying Waters Protected by the Clean Water Act," April 27, 2011, p. 2, on file with author.

[26] For information on the 111th Congress legislation, see CRS Report R41225, *Legislative Approaches to Defining "Waters of the United States"*, by Claudia Copeland.

[27] Nancy Sutley, Chair, Council on Environmental Quality, et al., letter to Senator Barbara Boxer, Chair, Senate Environment and Public Works Committee (and other congressional leaders), May 20, 2009, http://epw.senate.gov/ public/index.cfm?FuseAction=Majority.PressReleases &ContentRecord_id=64739ae3-802a-23ad-4c30-36fc58cc1014& Region_id=&Issue_id=.

[28] For information, see CRS Report R43076, *The 2013 Farm Bill: A Comparison of the Senate-Passed (S. 954) and House-Passed (H.R. 2642, H.R. 3102) Bills with Current Law*, coordinated by Ralph M. Chite.

[29] For additional information, see CRS Report R40763, *Agricultural Conservation: A Guide to Programs*, by Megan Stubbs.

[30] USDA issued regulations to implement these changes to the Wetlands Reserve Program in January 2009. See 74 *Federal Register* 2317 (January 15, 2009).

[31] For more information on these provisions, see CRS Report RL34557, *Conservation Provisions of the 2008 Farm Bill*, by Tadlock Cowan, Renée Johnson, and Megan Stubbs.

[32] For information, see CRS Report R43076, *The 2013 Farm Bill: A Comparison of the Senate-Passed (S. 954) and House-Passed (H.R. 2642, H.R. 3102) Bills with Current Law*, coordinated by Ralph M. Chite.

[33] See http://www.usace.army.mil/CECW/Documents/cecwo/reg/mou/foodsecurity_clean water act.pdf.

[34] See http://www.fws.gov/partners/viewPage=partners.

[35] For information, see http://www.fws.gov/coastal/coastalgrants/.

[36] U.S. Congress, House of Representatives, Committee on Transportation and Infrastructure, Subcommittee on Water Resources and Environment, *The Wetland Permitting Process: Is It Working Fairly?* Hearing 107-50, 107th Cong., 1st sess., October 3, 2001.

[37] Environmental Law Institute, *State Wetland Protection: Status, Trends & Model Approaches*, March 2008, p. 6. Hereinafter, ELI State Wetland Protection.

[38] See Association of State Wetland Managers, "State Wetland Program Summaries," at http://www.aswm.org/statesummaries.

[39] Twenty-two states rely on Section 401 as the sole form of state-level regulation, and 15 additional states rely on Section 401 as the primary form of state-level regulation but also have adopted laws that provide additional protection to certain wetland categories, such as coastal wetlands. ELI State Wetland Protection, p. 13.

[40] See Environmental Council of the States, "The States' Definitions of 'Waters of the State,'" February 2009, at http://www.ecos.org/section/publications.

[41] ELI State Wetland Protection, p. 17.

[42] B.R. Couvillion, J.A. Barras, and G.D. Steyer, et al., *Land Area Change in Coastal Louisiana from 1932 to 2010*, U.S. Georglocal Survey, Pamphlet to accompany U.S. Geological Survey Scientific Investigations Map 3164, June 2011, http://pubs.usgs.gov/sim/3164/downloads/SIM3164_Pamphlet.pdf.

[43] For information on this program, see CRS Report RS22467, *Coastal Wetlands Planning, Protection, and Restoration Act (CWPPRA): Effects of Hurricanes Katrina and Rita on Implementation*, by Jeffrey A. Zinn.

[44] U.S. Government Accountability Office, *Coastal Wetlands: Lessons Learned from Past Efforts in Louisiana Could Help Future Restoration and Protection*, GAO-08-130, 57 p.

[45] See http://www.coast2050.gov. For a more detailed discussion of the effects of the hurricanes on planning for wetland restoration, see CRS Report RS22276, *Coastal Louisiana Ecosystem Restoration After Hurricanes Katrina and Rita*, by Jeffrey A. Zinn.

[46] S. 3711 was attached to a broad tax relief measure that was enacted in December 2006 (H.R. 6111, P.L. 109-432). For additional information, see CRS Report RL33493, *Outer Continental Shelf: Debate Over Oil and Gas Leasing and Revenue Sharing*, by Marc Humphries.

[47] Dennis F. Whigham, Stephen W. Broome, and Curtis J. Richardson, et al., Statement of the Environmental Concerns Committee, Society of Wetland Scientists, "The Deepwater Horizon Disaster and Wetlands," http://www.sws.org/docs/ SWS_OilEffectsOn Wetlands.pdf.

[48] For additional information, see CRS Report R41311, *The Deepwater Horizon Oil Spill: Coastal Wetland and Wildlife Impacts and Response*, by M. Lynne Corn and Claudia Copeland.

[49] Thomas E. Dahl and Susan-Marie Stedman, *Status and Trends of Wetlands in the Coastal Watersheds of the Conterminous United States 2004-2009*, U.S. Department of the Interior, Fish and Wildlife Service, and National Ocean and Atmospheric Administration, National Marine Fisheries Service, October 2013, p. 37.

[50] National Academy of Sciences, National Research Council, *Compensating for Wetland Losses under the Clean Water Act* (Washington, DC: 2001), 267 pp.

[51] U.S. Government Accountability Office, *Wetlands Protection: Assessments Needed to Determine the Effectiveness of In-Lieu-Fee Mitigation*, GAO-01-325, 75 pp.

[52] U.S. Environmental Protection Agency and U.S. Army Corps of Engineers, "National Wetlands Mitigation Action Plan, December 24, 2002." See http://water 2003_07_10_wetlands_map1226withsign.pdf.

[53] U.S. Army Corps of Engineers and Environmental Protection Agency, "Compensatory Mitigation for Losses of Aquatic Resources, Final Rule," 73 *Federal Register* 19594, April 10, 2008.

[54] Information on compensatory mitigation can be found at http://waterwetlands mitigation_index.cfm.

[55] For more information on mitigation generally, and mitigation banks specifically, see Environmental Law Institute, *2005 Status Report on Compensatory Mitigation in the United States*, April 2006, 105 pp.

In: U.S. Wetlands
Editor: Harriet M. Hutson

ISBN: 978-1-63117-800-9
© 2014 Nova Science Publishers, Inc.

Chapter 2

THE WETLANDS COVERAGE OF THE CLEAN WATER ACT (CWA): *RAPANOS* AND BEYOND[*]

Robert Meltz and Claudia Copeland

SUMMARY

In 1985 and 2001, the Supreme Court grappled with issues as to the geographic scope of the wetlands permitting program in the federal Clean Water Act (CWA). In 2006, the Supreme Court rendered a third decision, *Rapanos v. United States*, on appeal from two Sixth Circuit rulings. The Sixth Circuit rulings offered the Court a chance to clarify the reach of CWA jurisdiction over wetlands adjacent only to *non*navigable tributaries of traditional navigable waters—including tributaries such as drainage ditches and canals that may flow intermittently. (Jurisdiction over wetlands adjacent to traditional navigable waters was established in the 1985 decision.)

The Court's decision provided little clarification, however, splitting 4-1-4. The four-Justice plurality decision, by Justice Scalia, said that the CWA covers only wetlands connected to relatively permanent bodies of water (streams, rivers, lakes) by a continuous surface connection. Justice Kennedy, writing alone, demanded a substantial nexus between the wetland and a traditional navigable water, using an ambiguous ecological

---

[*] This is an edited, reformatted and augmented version of Congressional Research Service Publication, No. RL33263, dated November 6, 2013.

test. Justice Stevens, for the four dissenters, would have upheld the existing broad reach of Corps of Engineers/EPA regulations.

Because no rationale commanded the support of a majority of the Justices, lower courts are extracting different rules of decision from *Rapanos* for resolving future cases. Corps/EPA guidance issued in December 2008 says that a wetland generally is jurisdictional if it satisfies either the plurality or Kennedy tests. In April 2011, the agencies proposed revised guidance intended to clarify whether waters are protected by the CWA, but this proposal was controversial. The ambiguity of the *Rapanos* decision and questions about the agencies' guidance have increased pressure on EPA and the Corps to initiate a rulemaking to promulgate new regulations. In September 2013, EPA and the Corps withdrew the controversial proposed guidance and submitted a draft rule to the Office of Management and Budget for review. The substance of the draft rule, and when it might be proposed, are unknown for now. There also has been pressure on Congress to provide legislative clarification. In the 111th Congress, legislation intended to do so was approved by a Senate committee, but no further legislative action occurred. Similar legislation was not introduced in the 112th Congress or so far in the 113th Congress. Instead, proposals to bar issuance of the Corps/EPA revised guidance and to narrow the regulatory scope of the CWA have been introduced.

The legal and policy questions associated with *Rapanos*—regarding the outer geographic limit of CWA jurisdiction and the consequences of restricting that scope—have challenged regulators, landowners and developers, and policymakers for 40 years. The answer may determine the reach of CWA regulatory authority not only for the wetlands permitting program but also for other CWA programs, since the CWA uses but one jurisdiction-defining phrase ("navigable waters") throughout the statute.

While regulators and the regulated community debate the legal dimensions of federal jurisdiction under the CWA, scientists contend that there are no discrete, scientifically supportable boundaries or criteria along the continuum of wetlands to separate them into meaningful ecological or hydrological compartments. Wetland scientists believe that all such waters are critical for protecting the integrity of waters, habitat, and wildlife downstream. Changes in the limits of federal jurisdiction highlight the role of states in protecting waters not addressed by federal law. From the states' perspective, federal programs provide a baseline for consistent, minimum standards to regulate wetlands and other waters. Most states are either reluctant or unable to take steps to protect non-jurisdictional waters through legislative or administrative action.

INTRODUCTION

In 2006, the Supreme Court decided *Rapanos v. United States*,[1] the most recent and well-known of three Supreme Court decisions wrestling with the question of which wetlands are covered by the wetlands permitting program in the Clean Water Act (CWA).[2] Since then, numerous decisions from the lower federal courts have sought to divine what criteria to draw from the fractured opinions in *Rapanos* as to which wetlands are "jurisdictional" (within the CWA's reach), and which are not. At the same time, the agencies charged with administering the wetlands permitting program, the U.S. Army Corps of Engineers and Environmental Protection Agency (EPA), have issued several guidance documents seeking to explain their view of their jurisdiction post-*Rapanos*.

This report provides background including the pre-*Rapanos* Supreme Court opinions, then moves on to *Rapanos* itself and the Corps/EPA guidance documents.

BACKGROUND

From the earliest days, Congress has grappled with where to set the line between federal and state authority over the nation's waterways. Typically, this debate occurred in the context of federal legislation restricting uses of waterways that impaired navigation and commerce. The phrase Congress often used to specify waterways over which the federal government had authority was "navigable waters of the United States."[3] This "navigable waters" concept proved an elastic one: in Supreme Court decisions from the early to mid-20th century, "navigability" underwent a substantial expansion "from waters in actual use to those which used to be navigable to those which by reasonable improvements could be made navigable to nonnavigable tributaries affecting navigable streams."[4]

Notwithstanding the Court's enlargement of "navigability," the Congress considering the legislation that became the CWA of 1972[5] felt that the term was too constricted to define the reach of a law whose purpose was not maintaining navigability, as in the past, but rather preventing pollution. Accordingly, Congress in the CWA retained the traditional term "navigable waters," but defined it to mean "waters of the United States"[6]—seemingly minimizing the constraint of navigability. The conference report said that the new phrase was intended to be given "the broadest possible constitutional interpretation."[7]

Among the provisions in the 1972 clean water legislation was Section 404,[8] which together with Section 301(a) requires persons wishing to discharge dredged or fill material into "navigable waters," as newly defined, to obtain a permit from the U.S. Army Corps of Engineers.[9] The Corps' initial response to Section 404 was to apply it solely to waters traditionally deemed navigable (which included few wetland areas), despite the broadening "waters of the United States" definition and conference report language. Under a 1975 court order,[10] however, the Corps issued new regulations that swept in a range of wetlands.[11] This broadening ushered in a debate, continuing today, as to *which* wetlands Congress meant to reach in the Section 404 permit program. At one time or another, the debate has occupied all three branches of the federal government.

Wetlands, with a variety of physical characteristics, are found throughout the country. They are known in different regions as swamps, marshes, fens, potholes, playa lakes, or bogs. Although these places can differ greatly, they all have distinctive vegetative assemblages because of the wetness of the soil. Some wetland areas may be continuously inundated by water, while other areas may not be flooded at all. In coastal areas, flooding may occur on a daily basis as tides rise and fall.

Riverside Bayview Homes

The Supreme Court's first foray into the Section 404 jurisdictional quagmire came in 1985, in *Riverside Bayview Homes, Inc. v. United States*.[12] There, the Court unanimously upheld as reasonable the Corps' extension of its Section 404 jurisdiction to "adjacent wetlands"—as one component of the agency's definition of "waters of the United States."[13] Under the Corps regulations, adjacent wetlands are wetlands adjacent to any non-wetland waterbody that constitutes a water of the United States—such as navigable bodies of water or interstate waters, or their tributaries. The Court reasoned that the water-quality objectives of the CWA were broad and sensitive to the fact that water moves in hydrologic cycles. Due to the frequent difficulties in defining where water ends and land begins, the Court could not say that the Corps' conclusion that adjacent wetlands are inseparably bound up with "waters of the United States" was unreasonable, particularly given the deference owed by courts to the Corps' and EPA's ecological expertise. Also persuasive was the fact that in considering the 1977 amendments to the CWA, Congress vigorously debated but ultimately rejected amendments that would have narrowed the Corps' asserted jurisdiction under Section 404.

SWANCC

In 2001, the Court returned to the geographic reach of Section 404. The decision in *Solid Waste Agency of Northern Cook County v. U.S. Army Corps of Engineers (SWANCC)*[14] directly involved the "isolated waters" component of the Corps' definition of "waters of the United States,"[15] rather than the "adjacent wetlands" component at issue in *Riverside Bayview Homes*. "Isolated waters," in CWA parlance (the regulations don't actually use the phrase), are waters that are not traditional navigable waters, are not interstate, are not tributaries of the foregoing, and are not hydrologically connected to navigable or interstate waters or their tributaries—but whose "use, degradation, or destruction [nonetheless] could affect interstate commerce."[16] Illustrative examples listed in the regulations include "intrastate lakes, rivers, streams (including intermittent streams), mudflats, sandflats, *wetlands*, sloughs, [or] prairie potholes"[17] with an interstate commerce nexus, or connection. The issue before the Court was whether "waters of the United States" is broad enough to embrace the Corps' assertion of jurisdiction over such "isolated waters" purely on the ground that they are or might be used by migratory birds that cross state lines—known as the Migratory Bird Rule.

In a 5-4 ruling, the Court held that the Migratory Bird Rule was not authorized by the CWA. The decision's rationale was much broader, however, appearing to preclude federal assertion of 404 jurisdiction over isolated, nonnavigable, intrastate waters on *any* basis—indeed, over wetlands not adjacent to "open water."[18] This disparity between the Court's holding and its rationale occasioned considerable litigation in the lower courts, the majority of which opted for a narrow reading of *SWANCC*, hence a broad reading of remaining Corps jurisdiction under Section 404. Such uncertainties as to the Corps' isolated waters jurisdiction after *SWANCC* focused attention on the alternative bases in Corps regulations for asserting 404 jurisdiction—such as the existence of "adjacent wetlands." Neither the Corps of Engineers nor EPA, however, has modified its Section 404 regulations since *SWANCC*.[19]

The new spotlight on the concept of "adjacent wetlands" became the backdrop for the Supreme Court's *Rapanos* decision, the Court's second encounter with this phrase after *Riverside Bayview Homes*.

RAPANOS

Rapanos was actually a consolidation of two cases, Rapanos and Carabell, on appeal from the Sixth Circuit. Though both cases involved issues as to what constitutes "adjacent wetlands," the issues in each are different.

The Sixth Circuit Decisions

Rapanos in the Sixth Circuit involved the Corps' assertion of 404 jurisdiction over a wetland adjacent to a tributary (man-made ditch) that ultimately flowed, miles later, into a traditional navigable water. As in Riverside Bayview, the issue was the Corps' jurisdiction under the "adjacent wetlands" component of its regulations defining "waters of the United States." In particular, plaintiffs argued that SWANCC did more than throw out the Migratory Bird Rule; it also barred Section 404 regulation of wetlands that do not physically abut a traditional navigable water.

In ruling that Section 404 reached the Rapanos's wetlands, the Sixth Circuit held that immediate adjacency of the wetland to a traditional navigable water is not required. Rather, what is needed is a "significant nexus"—a ubiquitous phrase in Section 404 court decisions lifted from SWANCC's explanation of Riverside Bayview[20]—between the wetlands and traditional navigable waters. "Significant nexus," in turn, can be satisfied by the presence of a "hydrological connection." Thus, the fact that the Rapanos's wetlands had surface water connections to nearby tributaries of traditional navigable waters was sufficient for Section 404 jurisdiction. Nor did it seem to matter to the court that the hydrological connection to traditional navigable waters was, for at least one of the Rapanos wetlands, distant—surface waters from this wetland flow into a man-made drain immediately north of the site, which empties into a creek, which flows into a navigable river. According to the record, this wetland is between 11 and 20 miles from the nearest navigable-in-fact water. In ruling that a surface water connection to a tributary of a navigable water was enough, the circuit aligned itself with the large majority of appellate courts to rule on this issue since SWANCC.

In its petition for certiorari to the Supreme Court, the Rapanoses asked whether the CWA's reach extends to nonnavigable wetlands "that do not even abut a navigable water."

Carabell in the Sixth Circuit involved the Corps' assertion of jurisdiction over a wetland adjacent to a tributary (man-made ditch) that ultimately flowed into traditional navigable waters—but the wetland was separated from the tributary by a manmade berm.

The Sixth Circuit held that "adjacent wetlands" jurisdiction existed under the Corps regulations, even though the wetland was separated from a tributary of "waters of the United States" by a four-foot-wide manmade berm that blocked immediate drainage of surface water from the parcel to the tributary.[21] The existence of the berm meant, critically, that unlike the wetlands in Rapanos, the wetlands here lacked any hydrological connection to navigable waters at all. Parenthetically, the fact that the "tributary" was merely a man-made ditch (which emptied into a creek, which flowed into a navigable lake) did not appear to be an issue in the case, as it was in Rapanos. Finally, the court endorsed the view of the majority of courts addressing the question that SWANCC spoke only to the Corps' "isolated waters" jurisdiction; it did not narrow the agency's "adjacent wetlands" authority involved here and broadly construed in Riverside Bayview.

In its petition for certiorari, the Carabells asked whether Section 404 extends to "wetlands that are hydrologically isolated from any of the 'waters of the United States.'"

The Supreme Court Decision

For many who had waited so long to have "waters of the United States" clarified, the *Rapanos* decision (addressing the Sixth Circuit decisions in both *Rapanos* and *Carabell*) was a disappointment. In three major opinions, the Court split 4-1-4 as to whether the Corps' assertions of 404 jurisdiction in the two cases before it comported with the CWA—that is, involved "waters of the United States." Justice Scalia wrote a four-Justice plurality opinion, ruling that the Corps had overreached and thus the Sixth Circuit decisions must be vacated and remanded for further proceedings applying the plurality's rule. Justice Kennedy, in a lone concurrence, also disagreed with the Corps' interpretation of the CWA, but would have applied a different approach than the plurality. He supplied the fifth vote supporting the vacation and remand, making that the judgment of the Court. (Five votes is a majority on the

Supreme Court.) Finally, Justice Stevens wrote a four-Justice dissent upholding the Corps' reading of its jurisdiction. Accordingly, he would have affirmed the decisions below.[22]

The problem is that no single rationale in these three opinions commands the support of a majority of the Justices. Thus, lower courts addressing challenges to Corps 404 jurisdiction since *Rapanos* have struggled with what rule of decision to extract from the decision. Does the Scalia plurality decision control? Or does the Kennedy concurrence provide the test? Or is satisfying either of these adequate to support jurisdiction?

Justice Scalia's plurality opinion asserts what is probably the narrowest view of 404 jurisdiction in the three major opinions, at least in most circumstances. His opening paragraphs set the tone by describing the substantial costs of applying for 404 permits, and the "immense expansion of federal regulation of land use that has occurred under the Clean Water Act."[23] This critical tone continues with the opinion's description of how the lower courts, "[e]ven after *SWANCC*," have continued to uphold the "sweeping" assertions of jurisdiction by the Corps over tributaries and adjacent wetlands.[24]

Justice Scalia goes on to construe "waters" in "waters of the United States" to mean only *relatively permanent, standing or flowing bodies of water*, such as streams, rivers, lakes, and other bodies of water "forming geographic features."[25] This definition leads him to exclude "channels containing merely intermittent or ephemeral flow."[26] Wetlands, our topic here, are included as "waters of the United States"—that is, are "adjacent" in the Corps' language—only when they have a *"continuous surface connection"* to bodies that are "waters of the United States" in their own right. By contrast, wetlands with only an intermittent, physically remote hydrological connection to "waters of the United States" are not covered by Section 404, according to the Scalia opinion.

Importantly, the plurality sought to calm concerns that a narrow reading of Section 404 would eviscerate other sections of the CWA, particularly the point-source permitting program under Section 402 that is the heart of the act. That section, the plurality explained, does not require that the point source discharge *directly* into a jurisdictional water. It is enough that the discharged pollutant is likely to ultimately be carried downstream to such a jurisdictional water. Thus, unlike with Section 404, discharges into non-covered waters could still be regulated.

Justice Kennedy's concurring opinion, in contrast to the absolute rules proposed by the plurality, offers a case-by-case test. He picks up on the "significant nexus" test used by the Sixth Circuit and many other courts—but while the lower courts defined significant nexus as having a *hydrological connection* with traditional navigable waters,[27] Justice Kennedy used an ambiguous *ecological* test.[28] A wetland, he declared, has the requisite significant nexus if, alone or in combination with similarly situated lands in the region, it significantly affects the chemical, physical, and biological integrity of traditional navigable waters.[29] These ecological functions include flood retention, pollutant trapping, and filtration. Under Kennedy's opinion, the waters that perform these functions may be intermittent or ephemeral, and they need not have a surface hydrological connection to other waters. When, in contrast, their effects on water quality are speculative or insubstantial, the wetland is beyond Section 404's reach.[30]

This formulation, Justice Kennedy explained, allows that when the Corps seeks to regulate wetlands adjacent to *navigable-in-fact* waters, adjacency is enough for jurisdiction. In contrast, for wetlands sought to be regulated based on adjacency to *non-navigable tributaries*, a significant nexus must be shown on a case-by-case basis. Importantly, however, the Justice did allow that the Corps might adopt regulations at some point declaring certain categories of wetlands to have a significant nexus per se, obviating the case-by-case approach for those wetlands.

Each of the foregoing views, the plurality's and Justice Kennedy's, rejects the hitherto prevailing view that any hydrological connection to a traditionally navigable water, no matter how distant, is sufficient for coverage. This "any hydrological connection" test had been a key element of the United States' assertions of "adjacent wetlands" jurisdiction.

The four dissenters found the Corps' assertion of jurisdiction reasonable in both cases. The Court's earlier decision in *Riverside Bayview*, the dissenters argue, was not confined to wetlands having continuous surface flow with traditional navigable waters or their tributaries. Rather it had endorsed jurisdiction over non-isolated wetlands generally, without case-by-case analysis. The plurality's concerns about the costs of applying for a permit, they continued, are more properly addressed to Congress, not to a court.

LEGAL ANALYSIS OF *RAPANOS*

The jurisdictional questions raised by Rapanos and Carabell presented the Supreme Court with a "perfect storm" of hot-button issues. First, there is the federalism matter: Where do CWA Section 404 and the Constitution's Commerce Clause draw the line between federal and state authority over wetlands? Second, there are property rights concerns. Some 75% of jurisdictional wetlands in the lower 48 states are on private property, with the result that protests from property owners denied Section 404 permits (or subjected to unacceptable conditions on same) are often heard— sometimes in the courts through Fifth Amendment takings suits. Third, Rapanos and Carabell have pervasive significance within the CWA itself, since "waters of the United States" governs not only the Section 404 wetlands permitting program, but also multiple other provisions and requirements of that law (see discussion below under "Policy Implications"). In addition, the Corps' broad reading of its jurisdiction created novel semantics (such as viewing dry arroyos as "waters," and manmade ditches as "tributaries") that Justices inclined to more literal readings of statutory language would have a hard time accepting.

It was not surprising in light of the above themes that the Justices split as they did: the four more "conservative" Justices rejecting the Corps' expansive view of its adjacent wetland jurisdiction, the four "liberal/moderates" upholding it, and Justice Kennedy coming down in between (as he often does) with a case-by-case test, at least until the Corps adopts new rules. The question, as noted earlier, is what rule of decision the lower courts will discern in Rapanos, with its absence of a majority rationale, for use in future cases. In practice, courts often look for common approaches supported by a majority of the Justices, looking both to the views of plurality Justices (supporting the judgment of the court in the case) and those of the dissenters (who do not support the judgment).

Thus far, lower courts applying Rapanos have drawn different tests from the decision, as was predicted based on its fractured nature. Nine of the thirteen federal circuits have ruled so far, an indication of the frequency with which CWA jurisdictional questions arise.[31] Two federal circuits held that the Kennedy "significant nexus" test alone controls;[32] two applied the Kennedy test but reserve for another day the question whether the plurality test as well is valid;[33] three accepted Justice Stevens's suggestion that a wetland satisfying either the Kennedy or plurality tests is jurisdictional;[34] and two avoided the issue altogether by finding that the Kennedy test and plurality test were both satisfied by the particular wetland in the case.[35] (See Figure A-1 in the Appendix to this report.) No circuit decision has opted for the plurality test alone. As the footnotes below show, the Supreme Court has declined to review every one of these circuit decisions where a petition for certiorari has been filed. The likely reason for these consistent denials is that with no change in the Justices since Rapanos that is likely to make a difference in their voting pattern, the Court may see little point to taking another case in the area.

District court decisions, at least the reported ones, seem to all follow either the Kennedy test alone or the Kennedy-or-plurality test view.[36] As with the appellate decisions, there appears to be no reported district-court decision squarely holding that the plurality test alone governs.[37]

To a considerable extent, the court decisions turn on how the courts read Supreme Court guidance on what rule of law may be inferred from decisions of the Court in which no rationale commands the support of five or more Justices. The United States, for its part, has consistently taken the Kennedy-or-plurality position in litigation, as it did in congressional testimony soon after the Rapanos decision[38] and in the Corps/EPA guidance on interpreting Rapanos (discussed below).

In the wake of Rapanos, several factors arguably put pressure on the Corps and EPA to do a rulemaking on the scope of "adjacent wetlands" permitting jurisdiction under the CWA (assuming Congress does not act). One is the fact that no fewer than three of the opinions in Rapanos urged the agencies to do so.[39] A second factor is the labor-intensive nature (and vagueness) of the Kennedy case-by-case approach, requiring empirical study of each wetland near a non-navigable tributary. The third factor is the divergence of the lower courts as to the rule to be applied after Rapanos. One can be confident, however, that anything the Corps and EPA promulgate will find its way into the courts. The agencies stated in guidance issued in 2008 that "further consideration of jurisdictional issues, including clarification and definition of key terminology, may be appropriate in the future, either through issuance of additional guidance or through rulemaking."[40]

All of the Rapanos opinions that mention SWANCC seem to accept, without discussion, that SWANCC eliminates jurisdictional coverage of all isolated, intrastate, nonnavigable waters—not just those isolated, intrastate, nonnavigable waters where the sole basis for asserting jurisdiction was the Migratory Bird Rule. Most lower court decisions to broach this issue had adopted the latter narrower reading of SWANCC. Thus, although only adjacent wetlands were directly involved in Rapanos, there may be impacts on the Corps' authority over isolated, intrastate, nonnavigable waters also.

Finally, although both petitions for certiorari in Rapanos raised the Commerce Clause issue, the decision in Rapanos, as expected, was on purely statutory grounds. The plurality, however, did assert that the Corps view of its adjacent wetlands jurisdiction "stretches the outer limits of Congress' commerce power,"[41] using this as one of several reasons for adopting a narrow reading of that jurisdiction. This plurality view is plainly relevant to congressional bills seeking to overturn SWANCC and Rapanos by amending the CWA to explicitly assert jurisdiction over waters to the fullest extent consistent with the Constitution (see "Legislative Consideration").

THE EPA/CORPS GUIDANCE ON *RAPANOS*

On December 2, 2008, EPA and the Corps of Engineers issued guidance to their field offices on how *Rapanos* should be interpreted in jurisdictional determinations, agency enforcement actions, and other agency actions. The guidance does not impose legally binding requirements on EPA or the Corps, and may not apply in a particular circumstance.

The Corps and EPA had previously issued other guidance, attempting to clarify the Court's rulings on the jurisdictional issues discussed here. Following the *Rapanos* ruling, the agencies first issued informal guidance in 2006; it was replaced by formal guidance in June 2007. The December 2008 guidance made limited changes to the 2007 guidance and supersedes it.[42] The 2008 revisions were made after review of public comments on the 2007 guidance and evaluation of the agencies' own implementation of the guidance. However, they noted in 2008, "The agencies will continue to monitor implementation of the Rapanos Guidance and, as we gain experience, consider appropriate opportunities to provide additional guidance or to initiate rulemaking."[43] This statement encouraged those who argue that revised regulations are needed to resolve lingering interpretive questions. Others contend that a legislative remedy is required. The potential for litigation to challenge the guidance itself is unclear.

The 2008 guidance adopts the Kennedy-test-or-plurality-test view, with the addition of agency interpretation of vague phrases in the Kennedy and plurality opinions. It has three parts, addressing waters that are (1) categorically within the scope of "waters of the United States"; (2) within "waters of the United States" or not, on a case-by-case basis; or (3) categorically outside the scope of "waters of the United States."

1) Waters categorically labeled "waters of the United States"—that is, without a case-by-case inquiry into whether there is a "significant nexus" with a traditional navigable water—are first, traditional navigable waters[44] and their adjacent wetlands. Under this test, the existence of a continuous surface connection, as demanded by the

plurality, but not Kennedy or the dissenters, is required to establish adjacency. Categorical "waters of the United States" also include non-navigable tributaries of traditional navigable waters, where such tributaries are "relatively permanent waters" (i.e., typically flowing year-round or at least seasonally) and adjacent wetlands with a continuous surface connection to such tributaries (not separated by uplands, berms, etc.). The 2008 guidance clarifies that a wetland is adjacent if it has an unbroken hydrologic connection to jurisdictional waters, or is separated from those waters by a berm or similar feature, or if it is in reasonably close proximity to a jurisdictional water.

2) Waterbodies that are "waters of the United States" on a case-by-case basis are those dependent on a finding of a "significant nexus" with a traditional navigable water, per the Kennedy concurrence. They include non-navigable tributaries that are not relatively permanent (such as intermittent and ephemeral streams) and their adjacent wetlands, and wetlands adjacent to but that do not directly abut a relatively permanent non-navigable tributary. The 2008 guidance states that, in making the site- and fact-specific analysis to determine "significant nexus," the agencies will evaluate hydrology (e.g., proximity to traditional navigable waters), ecologic factors (e.g., ability of wetlands to trap and filter pollutants or store flood waters), and flow characteristics (flow and functions of the tributary and adjacent wetlands). The purpose of these tests is to demonstrate a connection and the role of a tributary and any adjacent wetlands in protecting the chemical, physical, and biological integrity of downstream traditional navigable waters.

3) Waterbodies not generally considered "waters of the United States" are swales or erosional features (e.g., gullies) and ditches (including roadside ditches) excavated wholly in and draining only uplands, and that do not carry a relatively permanent flow of water. The agencies generally will not assert jurisdiction over these waterbodies.

To provide greater transparency of decisionmaking, the 2007 guidance required the Corps and EPA to be more thorough in documenting their jurisdictional determinations than in the past. To meet this requirement, which continues under the 2008 guidance, the Corps uses a standardized documentation form and posts results on District websites.[45] These steps respond to criticism, such as detailed in a GAO report, that Corps district offices have used differing practices in making jurisdictional determinations and that few districts made their documentation public.[46]

Overall, stakeholder groups, including industry, environmental advocates, and states, expressed disappointment or frustration with the 2007 guidance and the 2008 revision—some believing that it goes too far in narrowing protection of wetlands and U.S. waters, others believing that it does not go far enough. Generally, most agree that implementing the "significant nexus" test is especially difficult, because the guidance is complicated and vague. Industry groups said that because there are no clear guideposts on this key point, the guidance fails to provide the certainty desired by the regulated community. Environmentalists said that the "significant nexus" test in the guidance is more limited than the standard described by Justice Kennedy, because although his opinion recognizes the impact of losing wetlands or other small tributaries on large waters,[47] the guidance does not account for cumulative effects. In evaluating "significant nexus," the guidance focuses only on a tributary and wetlands adjacent to that tributary. The 2008 revised guidance did not modify the 2007 guidance with respect to evaluating "significant nexus." Overall, industry groups reportedly believe that the 2008 revisions provide modest improvement over the earlier guidance and could make some jurisdictional determinations easier, but environmental advocates assert that the guidance substantially limits the waters that will be protected by the Clean Water Act.[48]

One issue that has caused considerable confusion following the *Rapanos* ruling concerns CWA jurisdiction over wetlands not immediately adjacent to traditional navigable waters—including how jurisdiction will be applied in states within the Fourth, Seventh, Ninth, and Eleventh Circuits, where appellate courts have subsequently said that the Kennedy test alone is controlling. As noted, the 2008 guidance adds some clarification about determining adjacency, but continuing questions about this and other interpretive issues are possible.

Since the initial 2007 guidance was issued, the CWA permitting process has become more complex and is slower, according to many participants and observers. A revealing EPA memorandum in March 2008 reports that since July 2006 (shortly after *Rapanos* was decided), the *Rapanos* ruling or the 2007 guidance negatively affected approximately 500 enforcement cases, a "significant portion" of the CWA enforcement docket.[49] The breakdown identified in the EPA memo is 304 instances in which EPA regions decided not to pursue formal enforcement because of jurisdictional uncertainty, 147 instances where the enforcement priority of a case was lowered due to jurisdictional concerns, and 61 cases where lack of CWA jurisdiction has been asserted as an affirmative defense in an enforcement case. The memorandum goes on to say the greatest burden on the government results from "the implied presumption of non-jurisdiction [in the plurality test] for the most common types of waters in our country, intermittent and ephemeral tributaries to traditionally navigable waters and headwater wetlands. This presumptive exclusion can only be overcome by a resource-intensive 'significant nexus analysis' [the Kennedy test] described in the Guidance." The memorandum recommended "a few targeted revisions" to the guidance that OECA believes would address these issues, while remaining consistent with the *Rapanos* decision. For example, it recommended revising the guidance to incorporate Justice Kennedy's suggestions that, when evaluating jurisdiction, it is appropriate to consider wetlands either alone or in combination with other similarly situated lands in the region. The 2008 revised guidance did not address this recommendation.

Echoing the EPA memorandum, a Corps official stated at a May 2008 conference that making jurisdictional determinations is 8 to 10 times more resource-intensive for Corps staff who must consider a multitude of factors to determine what constitutes a "significant nexus." Representatives of developers and environmental advocates concurred that the joint guidance exacerbates permitting delays.[50] Concern about this reported impact on CWA enforcement drew the attention of two House committee chairmen in 2008. Their staffs reviewed a large number of EPA and Corps documents and concluded that there had been a significant decline in CWA inspections, investigations, and enforcement actions since the *Rapanos* ruling and the 2007 guidance.[51]

2011 Proposed Revised Guidance

At the end of 2010, the Obama Administration weighed into the CWA jurisdiction debate as EPA and the Corps drafted new joint agency guidance to clarify regulatory jurisdiction over U.S. waters and wetlands and to replace the agencies' 2008 guidance.

The document underwent several months of interagency review before release in April 2011 in the form of proposed revised guidance, subject to public comment until July 1, 2011. In releasing the proposal, the agencies said that once finalized, the revised guidance would supersede the 2008 guidance; until then, the existing guidance remains in effect. Also, the agencies said that after issuance of final guidance, they expected to propose revisions to regulations to further clarify which waters are subject to CWA jurisdiction, consistent with the Supreme Court's rulings.[52] As noted previously, three of the *Rapanos* opinions had argued for such a rulemaking.[53]

Status of the 2011 Proposed Guidance

As described below (see "Policy Implications"), in September 2013, the 2011 proposed guidance was withdrawn, prior to issuance in final form, and EPA and the Corps drafted a rule to clarify regulatory jurisdiction of the CWA. The draft rule is undergoing interagency review at the Office of Management and Budget (OMB). The following portion of this report discusses the proposed guidance. Although it has now been withdrawn, interest in the 2011 proposal remains high, particularly as an indication of policy that may be contained in the draft rule.

Like the 2008 guidance, the 2011 revisions proposed to adopt the Kennedy-test-or-plurality-test view of *Rapanos*. However, the agencies believed that a wider evaluation of jurisdiction is possible than the 2008 guidance suggests, stating, "after careful review of these opinions, the agencies concluded that previous guidance did not make full use of the authority provided by the CWA to include waters within the scope of the Act, as interpreted by the Court."[54] EPA and the Corps acknowledged that, compared with the existing guidance, the proposed revisions were likely to increase the number of waters identified as protected by the CWA.

> The agencies expect, based on relevant science and recent field experience, that under the understandings stated in this draft guidance, the extent of waters over which the agencies assert jurisdiction under the CWA will increase compared to the extent of waters over which jurisdiction has been asserted under existing guidance, though certainly not to the full extent that it was typically asserted prior to the Supreme Court decisions in *SWANCC* and *Rapanos*.[55]

EPA and Corps officials believed that the likely increase in jurisdictional waters would occur because, in their view, the existing guidance under-protects waters and has created uncertainty about many gray areas of jurisdiction, which the revised guidance was intended to clarify. Officials believed that additional acreage likely to be jurisdictional would not be large, but they did not estimate how much expansion would occur. Certain types of aquatic resources were likely to benefit, because they generally are non-jurisdictional under the existing guidance (for example, some prairie potholes). Although there still would be need for case-by-case determination of "significant nexus" waters (i.e., to demonstrate potential hydrologic or ecological connections to jurisdictional waters), the proposed revisions were intended to make such evaluations clearer.

The proposed revisions were built on the existing guidance with modifications that the agencies believed were consistent with the CWA, the Court's rulings, and science. According to EPA and the Corps, the guidance was focused on protection of smaller waters that feed into larger ones, to keep downstream water safe from upstream pollutants. The focus was also on reaffirming protection for wetlands that filter pollution and store water in order to help keep communities safe from floods. Two key changes in the proposed revisions, discussed below, were (1) inclusion of all interstate waters in those that are categorically labeled "waters of the United States," and (2) an expanded characterization of "adjacency" for waters and wetlands that must demonstrate a significant nexus to a jurisdictional water.

Like the existing guidance, the 2011 proposed revisions interpreted phrases in the plurality and Kennedy opinions in order to clarify waters that (1) are categorically within the scope of "waters of the United States;" (2) are "waters of the United States" if they meet a case-by-case test of "significant nexus" to a jurisdictional water; or (3) are generally outside the scope of "waters of the United States."

(1) Waters categorically labeled "waters of the United States" were those that are traditional navigable waters, wetlands adjacent to traditionally navigable waters or interstate waters, non-navigable tributaries of traditional navigable waters or interstate waters that are relatively permanent (i.e., they meet the plurality standard), and wetlands that directly abut relatively permanent waters. Existing regulations define "adjacent" to mean bordering, contiguous, or neighboring. Regarding categorically protected waters, the proposed guidance made two changes compared to the existing guidance: (a) all interstate waters were considered categorically jurisdictional, whereas under the existing guidance, interstate waters are not mentioned and their treatment is unclear, and (b) the standard for non-navigable tributaries that are "relatively permanent" was modified. Under the 2008 guidance, "relatively permanent" means a water that typically flows year-round or has continuous flow at least seasonally (meaning, e.g., typically for three months). Under the proposed revisions, seasonal flow making a water relatively permanent meant

that the water typically flows during the wet season of the region in which the water is located.

(2) Waters over which the agencies generally would assert jurisdiction were those that are not categorically jurisdictional but have "significant nexus" to a jurisdictional water. Under the Kennedy standard in *Rapanos*, waters have a "significant nexus" if, either alone or in combination with similarly situated waters in the region, they significantly affect the chemical, physical, or biological integrity of traditional navigable waters. The existing guidance says that the agencies will assert jurisdiction over non-navigable tributaries that are not relatively permanent and their adjacent wetlands, and over wetlands adjacent to but not directly abutting a relatively permanent tributary based on a fact-specific evaluation of significant nexus.

The 2008 guidance defines tributaries for the purpose of the "significant nexus" determination as a single stream reach of the same order.[56] It then limits "similarly situated" waters (described in the Kennedy standard) to a single stream reach or to wetlands that are adjacent to the same single stream reach. Aggregation of streams or stream reaches is not allowed under the 2008 guidance, and jurisdiction for tributaries under the "significant nexus" test is considered in isolation, that is, not together with other waters in the area.

The 2011 proposed guidance was broader and would allow watershed-wide aggregation for similarly situated wetlands and tributaries. Under the proposed guidance, a tributary would be jurisdictional if it is a tributary to a navigable-in-fact water or an interstate water and, alone or in combination with other tributaries in the watershed, has a significant nexus to such water. It essentially defined tributaries as those features that flow into another jurisdictional water and have a bed (the bottom of the channel) and bank, which is generally indicated by an ordinary high water mark.

Under the 2008 guidance, in evaluating significant nexus, EPA and the Corps can consider the flow characteristics (volume, duration, and frequency of flow) and ecological factors such as potential of the tributary to carry pollutants downstream to a traditional navigable water, or the potential of wetlands to trap and filter pollutants. The 2011 proposed guidance would allow consideration of other factors such as maintenance of habitat that provides spawning areas for species in downstream waters. It stated that all wetlands within a wetland mosaic (a landscape of wetlands and non-wetlands) should ordinarily be considered collectively when determining adjacency. As a result, wetlands likely to be considered jurisdictional under the revised guidance would be those that are adjacent to a newly identified traditional navigable water, or those that are adjacent to other jurisdictional non-wetland waters based on a more expansive "significant nexus" determination of adjacency.

Under the proposed revisions, so-called "other waters"[57] would require case-by-case evaluation and would be divided into two categories. First, physically proximate (see above) "other waters" that would satisfy the regulatory definition of "adjacent" if they were wetlands would be evaluated in the same manner as adjacent wetlands to determine significant nexus. Second, non-physically proximate "other waters" that do not meet the definition of "adjacent" in the proposed guidance (e.g., there is no demonstrable ecological interconnection to a jurisdictional waterbody) cannot be easily evaluated for significant nexus to jurisdictional waters. Thus the revised guidance proposed to continue the current practice of referring such waters to agency Headquarters for evaluation.

(3) Regarding waters that generally are not waters of the United States and over which the agencies will not assert jurisdiction, the existing guidance and 2011 proposed revisions were similar. Both would exclude geographic features such as swales, gullies, and ditches that are not tributaries or wetlands. The revised guidance also specifically identified a number of other examples of non-jurisdictional waters (e.g., waters that are excluded from coverage by statute or existing regulation; artificial lakes or ponds that are created by excavating and/or diking dry land and are used exclusively for purposes such as stock watering or irrigation; and artificially irrigated areas that would revert to upland should irrigation cease).

Release of the proposed revisions was accompanied by a preliminary economic analysis document prepared by EPA. It examined a range of indirect costs and benefits resulting from implementing guidance to clarify the scope of CWA jurisdiction. The analysis indicated that the majority of costs would result from incremental permitting costs and mitigation expenses incurred by entities seeking CWA Section 404 permits. Economic benefits also would result from water quality improvements such as protecting additional small streams and wetlands. EPA's analysis estimated that the benefits of implementing the proposed guidance would range from $162 million to $328 million annually, while the incremental costs would be between $87 million and $171 million. EPA acknowledged that valuing the benefits of the new guidance to wetlands and other waters poses many challenges, and the precision and accuracy of results were highly uncertain; nevertheless, EPA concluded that the analysis suggested that benefits were likely to justify costs.[58] Preparing an economic analysis to accompany the proposed guidance was a somewhat unusual step for EPA, likely reflecting anticipated significant interest in the new guidance. Economic analyses are typically developed to support formal regulatory proposals, rather than non-binding agency guidance.

Critical reaction to the proposed revisions began even before release of the document in April 2011. While the guidance was being developed and reviewed internally, press reports suggested that it would substantially increase waters subject to CWA regulation, raising concern among industry and some other groups that was confirmed when the proposed guidance was released.[59] Industry criticism focused on two issues: (1) the revised guidance would broaden the number and kinds of waters subject to regulation, in their view beyond what the CWA and the Supreme Court's rulings permit; and (2) government was attempting to effect policy change through nonbinding guidance that generally is not reviewable by courts. EPA and Corps officials responded that the guidance would not extend federal protection to any waters not historically protected under the Clean Water Act and would be fully consistent with the law, including decisions of the Supreme Court. Most state and local officials were supportive of clarifying the scope of CWA regulated waters, but some were concerned that expanding the CWA's scope could impose costs on states and localities as their own actions (e.g., transportation projects) become subject to new requirements. Environmental advocacy groups welcomed the new guidance, which would more clearly define U.S. waters that are subject to CWA protections but would not, they said, expand the reach of the law.

The proposed guidance drew congressional attention, as well, both before and after its release. Some Members wrote letters supporting issuance of new guidance to address the confusion resulting from the Supreme Court's rulings.[60] Others criticized the revised guidance as going beyond clarification and thus amounting to a *de facto* rule, instead of advisory guidelines.[61]

Legislative provisions to prohibit the agencies from funding activities related to revising the guidance were included in several appropriations bills in the 112th Congress, but none of these provisions was enacted. Similar provisions in 113th Congress appropriations bills also have not been enacted. Interest in legislation concerning the guidance also is evident with bills in the 113th Congress to prevent the agencies from finalizing the 2011 draft guidance (S. 861, S. 1006, S. 1514, and H.R. 1829). Other legislation would amend the CWA with a narrow definition of waters that are subject to the act's jurisdiction (S. 890 and H.R. 3377).

POLICY IMPLICATIONS

As with the legal questions, the policy questions associated with the Supreme Court cases—what *should be* the outer limit of CWA regulatory jurisdiction and what are the consequences of restricting that jurisdiction—also have challenged regulators, landowners and developers, and policymakers since passage of the act in 1972. The act prohibits the discharge of dredged or fill material into navigable waters without a permit, and it also prohibits discharges of pollutants from any point source to navigable waters without a permit. Disputes have centered on whether wetlands and other waters are "navigable waters," a legal term of art. The answer to this question is important, because it may determine the extent of federal CWA regulatory authority not only for the Section 404 program, but also for purposes of implementing other CWA programs. Critics of the Section 404 regulatory program, such as land developers and agriculture interests, argue that the Corps' wetlands program has gradually and illegally expanded its asserted jurisdiction since 1972. They want the Corps and EPA to give up jurisdiction over most non-navigable tributaries and allow other federal and state programs to fill whatever gap is created. Waters that are jurisdictional are subject to the multiple regulatory requirements of the CWA: standards, discharge limitations, permits, and enforcement. Non-jurisdictional waters, in contrast, do not have the federal legal protection of those requirements. The act has one definition of "navigable waters" that applies to the entire law. The definition applies to federal prohibition on discharges of pollutants (§301), requirements to obtain a permit prior to discharge (§§402 and 404), water quality standards and measures to attain them (§303), oil spill liability and oil spill prevention and control measures (§311), certification that federally permitted activities comply with state water quality standards (§401), and enforcement (§309). It impacts the Oil Pollution Act and other environmental laws, as well. For example, the reach of the Endangered Species Act (ESA) is affected, because that act's requirement for consultation by federal agencies over impacts on threatened or endangered species is triggered through the issuance of federal permits.[62] Thus, by removing the need for a CWA permit, a non-jurisdictional determination would eliminate ESA consultation, as well. As discussed above, the Scalia opinion in *Rapanos* concluded that a narrow interpretation of the Corps' 404 jurisdiction would not impact these other provisions, but many observers contend that the question is not fully resolved. EPA said that it might issue additional guidance concerning the effect of *Rapanos* on other CWA programs that use the common "waters of the United States" definition, but it has not done so. In March 2008, EPA officials reportedly asked states to assist in developing guidance to govern CWA jurisdiction decisions under Section 402, because of continuing uncertainty on the law's scope, especially in western states that have a preponderance of intermittent and ephemeral streams.[63] *SWANCC* found invalid the assertion of CWA jurisdiction over isolated, non-navigable intrastate waters solely on the basis of their use (or potential use) as habitat by migratory birds. Most of the post-*SWANCC* cases have instead addressed tributaries and adjacent wetlands, asking which of these have the "significant nexus" to navigable waters that *SWANCC* was interpreted to say is necessary to establish federal jurisdiction.

Wetlands are an important part of the total aquatic ecosystem, with many recognized functions and values, including water storage (mitigating the effects of floods and droughts), water purification and filtering, recreation, habitat for plants and animals, food production, and open space and aesthetic values. Functional values, both ecological and economic, at each wetland depend on its location, size, and relationship to adjacent land and water areas. To the layman, many of these values are more obvious for wetlands adjacent to large rivers and streams than they are for wetlands and small streams that are isolated in the landscape from other waters. Many of the functions and values of wetlands have been recognized only recently. Historically, many federal programs encouraged wetlands to be drained or altered because they were seen as having little value. Even today, while more federal laws either encourage wetland protection or regulate their modification, pressure exists to modify, drain, or develop wetlands for uses that some see as more economically beneficial.

While regulators and the regulated community debate the legal dimensions of federal jurisdiction, scientists contend that there are no discrete, scientifically supportable boundaries or criteria along the continuum of waters/wetlands to separate them into meaningful ecological or hydrological compartments. Numerous scientific studies define and describe the importance of the functions and values of wetlands, in support of their significant nexus to navigable waters.[64] In all but some very narrow instances, scientists say, terms such as "isolated waters" and "adjacent wetlands" are artificial legal or regulatory constructs, not valid scientific classifications. From this perspective, even waters and wetlands that lack a direct surface connection to navigable waters or that only flow intermittently are connected to the larger aquatic ecosystem via subsurface or overflow hydrologic connections. Wetland scientists believe that all such waters/wetlands are critical for protecting the integrity of waters, habitat, and wildlife downstream.

In *SWANCC*, the Supreme Court did not draw a bright line for purposes of determining the limits of federal jurisdiction (many wetland scientists do not believe that a bright line is possible, in any case). While the ruling reduced federal jurisdiction over some previously regulated wetlands, even more than a decade later it remains difficult to determine the precise effect of that decision. Many affected interests (states and the regulated community) contend that the 2003 guidance from the Corps and EPA did not adequately define the scope of regulated areas and wetlands affected by *SWANCC* and subsequent court rulings.[65] The Rapanoses and the Carabells had hoped that the Supreme Court would clarify the jurisdiction issue and that the Court would further narrow the program's geographic reach. Other interest groups disagreed with the petitioners' views on the issues, but also had hoped for clarity. Most say that the 4-1-4 ruling, in which the three main opinions did not agree on what constitutes "waters of the United States," did not bring the desired clarity of meaning in legal and policy terms. Estimates of the types of wetlands and amounts of acreage affected by *SWANCC, Rapanos*, and subsequent lower court rulings depend on interpretation of the cases and on assumptions about defining key terms such as "adjacent," "tributary," and "significant nexus." Because in its regulations before *SWANCC* the Corps had broadly defined "waters of the United States," including those encompassed by the Migratory Bird Rule, nearly all U.S. wetlands and waters were subject to CWA jurisdiction, since practically all are used to a greater or lesser extent by migratory birds.[66] Depending on how key terms are defined, reduced federal jurisdiction could affect very small or very large categories of waters and wetlands. Reflecting the uncertainties about how broadly or narrowly *SWANCC* would be interpreted, one estimate made after that decision found that the possible changes in jurisdiction could range from 20% to 80% of the nation's total estimated 100 million acres of wetlands.[67] Following the *Rapanos* decision, concern was expressed particularly about that ruling's impacts in arid and semi-arid western states to exclude intermittent or ephemeral streams and adjacent wetlands and riparian areas from CWA jurisdiction. A reduction in CWA jurisdiction affects implementation of the 404 and possibly other CWA programs. Early in 2006, EPA estimated conservatively that the extent of non-navigable tributaries and adjacent wetlands that could be affected by the narrow reading of the Clean Water Act that was advocated by the *Rapanos* and *Carabell* petitioners was up to 59% of the total length of streams in the United States, excluding Alaska. EPA also estimated that 34% of industrial and municipal dischargers that are subject to CWA Section 402 permits are located on these stream segments and that public drinking water systems which use intakes on these segments provide drinking water to over 110 million people.[68] Because there is no national database of non-navigable tributaries, EPA analyzed surrogate data on the linear extent of intermittent/ephemeral streams and stream segments that lie at the head of tributary systems and have no other streams flowing into them. Some estimate that the smallest, or headwater, first- and second-order streams represent more than 75% of the nation's stream network. These streams, if left unprotected by expansive interpretation of the Court's rulings, are at risk from a variety of polluting activities due to urbanization, construction, and channelization for flood control purposes.[69] As noted, the uncertainties resulting from the *Rapanos* decision led to widespread anticipation that the Corps and EPA would take administrative action to clarify how they interpret the ruling and its impact on waters that are protected by the Clean Water Act. Corps and EPA officials testified before a Senate subcommittee in August 2006 that the agencies were working on substantive interpretive guidance to clarify CWA jurisdiction in light of the decision[70]—the guidance that was eventually released in June 2007 and was revised in December 2008. While many observers acknowledged that guidance is useful, some argued that the Corps must initiate a rulemaking to revise its regulations—especially since three Justices in some fashion suggested one. This view is, in fact, widely held by many diverse stakeholders—environmental groups, industry, and states—who disagree on the substance of the guidance proposed by EPA and the Corps in April 2011. The Administration's schedule for either final guidance or new regulations was uncertain until recently. Final guidance was submitted to the White House Office of Management and Budget (OMB) for review in February 2012, but it was not released. In September 2013, EPA and the Corps announced that the guidance document had been withdrawn (a listing for the guidance was removed from OMB's list at http://www.reginfo.gov of all regulations under active review), and at the same time, the agencies submitted a draft regulation to OMB for review. The substance of this proposal, and when it might be proposed, are unknown for now. New regulations may clarify many current questions, but they are unlikely to please all of the competing interests, as one environmental advocate observed.

> However, a rulemaking would only benefit wetlands if it did not reduce the jurisdiction offered by current regulations and if the Administration remained faithful to sound science. If politics were to trump science in the rulemaking process, the likelihood of such a protective rule would not be promising. Also, rules are subject to

legal challenge and can be tied up in court for years before they are implemented.[71]

Also in September, EPA released a draft report that reviews and synthesizes the peer-reviewed scientific literature on the connectivity or isolation of streams and wetlands relative to large water bodies such as rivers, lakes, estuaries, and oceans. The purpose of the review, according to EPA, is to summarize current understanding about these connections, the factors that influence them, and mechanisms by which connected waters affect the function or condition of downstream waters.[72] The focus of the report is on small or temporary non-tidal streams, wetlands, and open waters. Based on the reviewed literature, it makes certain findings.

- All tributary streams, including perennial, intermittent, and ephemeral streams, are physically, chemically, and biologically connected to downstream rivers.
- Wetlands and open waters in riparian areas and floodplains also are physically, chemically, and biologically connected with rivers and serve an important role in the integrity of downstream waters. In these types of wetlands, water-borne materials can be transported from the wetland to the river network and vice versa (e.g., water from a stream flows into and affects the wetland).
- Wetlands and open waters where water only flows from the wetland to a river network, such as many prairie potholes, vernal pools, and playa lakes, occur on a gradient of connectivity, making it difficult to generalize about their effects on downstream waters from the currently available literature.

EPA has asked its Science Advisory Board (SAB) to review the draft report and to comment on whether its conclusions and findings are supported by the available science. The SAB review, which includes public meetings in mid-December, will presumably influence a final report at some future time. Earlier versions of the science report, in 2011, received interagency and external peer review, but were not released for public or SAB comment. The draft report is not intended as a policy document—it does not reference either the Scalia plurality or Kennedy tests in *Rapanos*, nor does it address legal standards for CWA jurisdiction. Nevertheless, EPA and the Corps say that the draft rule now at OMB takes into consideration the latest peer-reviewed science reflected in the draft science report. The report, when finalized, will provide a scientific basis needed to clarify CWA jurisdiction, according to EPA.[73] Some stakeholders have expressed concern that the scientific study could allow the agencies to assert jurisdiction in a blanket fashion over ephemeral and intermittent streams, rather than subjecting them to case-by-case determination of a "significant nexus" to downstream navigable waters.[74]

Filling the Gaps

Whatever gaps in wetland regulation result from reduced federal jurisdiction arguably could be filled, at least in part, by other federal or state and local programs and actions. For example, some assert that wetland restoration and creation programs, such as the Wetlands Reserve Program and the Coastal Wetlands Restoration Program, or private conservation efforts can provide protection, even if the wetland is no longer jurisdictional under federal law.[75] However, others respond that such programs are likely to be incomplete in filling gaps, since they apply primarily to rural areas and do not apply to the one-third of the nation's lands in federal ownership. Moreover, they were never intended to be a seamless group that would fill all possible gaps.

SWANCC, Rapanos, and the subsequent lower court decisions also highlight the role of states in protecting waters not addressed by federal law. From the states' perspective, the federal Section 404 program provides the basis for a consistent national approach to wetlands protection. But if a larger portion of wetlands are no longer jurisdictional, they say, it can be argued that the Section 404 program no longer provides a baseline for consistent, minimum standards to regulate wetlands. None of these court rulings prevents states from protecting non-jurisdictional waters through legislative or administrative action, but few states have done so. Prior to *SWANCC*, 15 states had programs that regulate isolated freshwater wetlands to some degree, but state officials acknowledge that these programs vary substantially from some that are comprehensive in scope to others that are limited by wetland size or have exemptions for agriculture and other activities.[76] Since 2001, a few states have passed new legislation or updated water quality regulations; the issue remains under consideration in several states, where competing proposals that are viewed by some as strengthening and by others as weakening wetland protection have been debated.[77]

Although some states have authorities to regulate waters of their state, their ability to regulate effectively may be compromised, because state rules often are tied to federal definitions. The gap produced by reduced federal jurisdiction is most evident in the 32 states that have no independent wetlands programs and that typically have relied on CWA Section 401 water quality certification procedures to protect wetlands. Pursuant to Section 401, applicants for a federal permit must obtain a state certification that the project will comply with state water quality standards. Consequently, by conditioning certification, states have the ability to affect the federal permit and to exercise some regulatory control over wetlands without the expense of establishing independent state programs. However, as described previously, diminished CWA jurisdiction which affects the Section 404 program also limits the reach of other CWA programs, including Section 401.

Analysts familiar with the political and fiscal environments of states believe that most states are either reluctant or unable "to step boldly into the breach in federal wetlands protection.... The Corps and the U.S. Environmental Protection Agency, not to mention Congress, have little cause to rely on the notion that states will effectively backstop federal protection for isolated wetlands."[78] Many states are barred from enacting laws more stringent than federal rules, or are reluctant to take action, due to budgetary and resource concerns, as well as apprehension that regulation will be judged to involve "taking" of private property and require compensation.

Legislative Consideration

Some argue that what is needed—regardless of interpretive guidance or rulemaking by the Corps and EPA—is legislative action to affirm Congress's intention regarding CWA jurisdiction. Others contend that, although the *Rapanos* decision did not resolve the issues, it also did not substantially affect Congress's willingness or interest in acting on issues that have been pending for several years without congressional action. Related to this is the view that, because the current questions are highly technical in nature, a simple fix may not address the problem, or may create others, such as impacting rights that the CWA reserves to states. In the 109th Congress, bills were introduced to address the CWA jurisdictional issues in different ways, but Congress took no action. One proposal (the Clean Water Authority Restoration Act of 2005) would have provided a broad statutory definition of "waters of the United States"; would have clarified that the CWA is intended to protect U.S. waters from pollution, not just maintain their navigability; and would have included a set of findings to assert constitutional authority over waters and wetlands. Other legislation intended to restrict regulatory jurisdiction also was introduced (the

Federal Wetlands Jurisdiction Act of 2005). It would have narrowed the statutory definition of "navigable waters" and defined certain isolated wetlands that are not adjacent to navigable waters, or non-navigable tributaries and other areas (such as waters connected to jurisdictional waters by ephemeral waters, ditches or pipelines), as not being subject to federal regulatory jurisdiction.

Legislation similar to the Clean Water Authority Restoration Act of 2005 was introduced in the 110th Congress (H.R. 2421 and S. 1870, a slightly different bill). The House Transportation and Infrastructure Committee held hearings on H.R. 2421 and related jurisdictional issues in July 2007, and a third hearing in April 2008. The Senate Environment and Public Works Committee held a non-legislative hearing on issues related to the *Rapanos* and *SWANCC* rulings in December 2007, and a legislative hearing on S. 1870 in April 2008.

Proponents of legislation contend that Congress must clarify the important issues left unsettled by the Supreme Court's 2001 and 2006 rulings and by the Corps/EPA guidance. Bill sponsors argued that the legislation would "reaffirm" what Congress intended when the CWA was enacted in 1972 and what EPA and the Corps had subsequently been practicing until recently, in terms of CWA jurisdiction. However, critics asserted that by making *activities* that affect waters of the United States (in addition to discharges) subject to the CWA's jurisdiction, the legislation would expand federal authority, and thus would have consequences that are likely to increase confusion, rather than settle it. Critics questioned the constitutionality of the bill, arguing that, by including all non-navigable waters in the jurisdiction of the CWA, it would exceed the limits of Congress's authority under the Commerce Clause. Supporters contended that the legislation is properly grounded in Congress's commerce power. The Bush Administration did not take a position on any legislation to clarify the scope of "waters of the United States" protected under the CWA. Congressional attention resumed in the 111th Congress, especially after statements by Obama Administration officials supporting the need for legislative clarification of these issues. In May 2009, the heads of EPA, the Corps, the Department of Agriculture, the Department of the Interior, and the Council on Environmental Quality jointly wrote to congressional leaders to identify certain principles that might help guide legislative and other actions: Broadly protect the nation's waters; make the definition of covered waters predictable and manageable; promote consistency between CWA and agricultural wetlands programs; and recognize long-standing practices, such as exemptions now in effect only through regulations or guidance.[79] A modified version of legislation from the previous Congress was introduced in the Senate (S. 787, the Clean Water Restoration Act), and in June 2009, the Senate Environment and Public Works approved it with an amendment in the nature of a full substitute to the bill as introduced. As approved by the committee, S. 787 would have deleted "navigable waters" from the CWA and use "waters of the United States" directly to define jurisdiction. It defined "waters of the United States" by a rewritten version of the regulatory definition in use by EPA and the Corps—

> The term "waters of the United States" means all waters subject to the ebb and flow of the tide, the territorial seas, and all interstate and intrastate waters including lakes, rivers, streams (including intermittent streams), mudflats, sandflats, wetlands, sloughs, prairie potholes, wet meadows, playa lakes, and natural ponds, all tributaries of any of the above waters, and all impoundments of the foregoing.

In response to prior criticism, the definition did not encompass *activities* that affect waters of the United States (see above). The bill as reported also instructed that "waters of the United States" be construed consistently with (1) how EPA and the Corps interpreted and applied "waters of the United States" prior to January 9, 2001, the day before *SWANCC* was decided, and (2) Congress's constitutional authority. The bill would have excluded, as in current EPA-Corps regulations, prior converted cropland and waste treatment systems. It also included a savings section that referenced without paraphrasing eight provisions in CWA §§402(l) and 404(f) which exempt certain types of discharges from CWA permits, such as discharges from normal farming activities, and discharges from maintenance of drainage ditches. The full Senate did not take up the bill.[80] Legislation similar to the bills in the 111th Congress was not re-introduced in the 112th Congress or so far in the 113th Congress, while, as described above, bills have been introduced to block EPA and the Corps from issuing revised "waters of the United States" guidance, as well as other bills to narrow the statutory definition of waters that are subject to CWA jurisdiction. In light of the widely differing views of proponents and opponents, future prospects for legislation on the geographic scope of CWA jurisdiction are highly uncertain. One difficulty of legislating changes to the CWA in order to specify which waters and wetlands are subject to the act's jurisdiction results from the fact that the complex scientific questions about such areas are not easily amenable to precise resolution in law. Debates over whether and how to revise the act highlight the challenges of trying to use the law to do so.

APPENDIX. WHICH *RAPANOS* TEST GOVERNS?

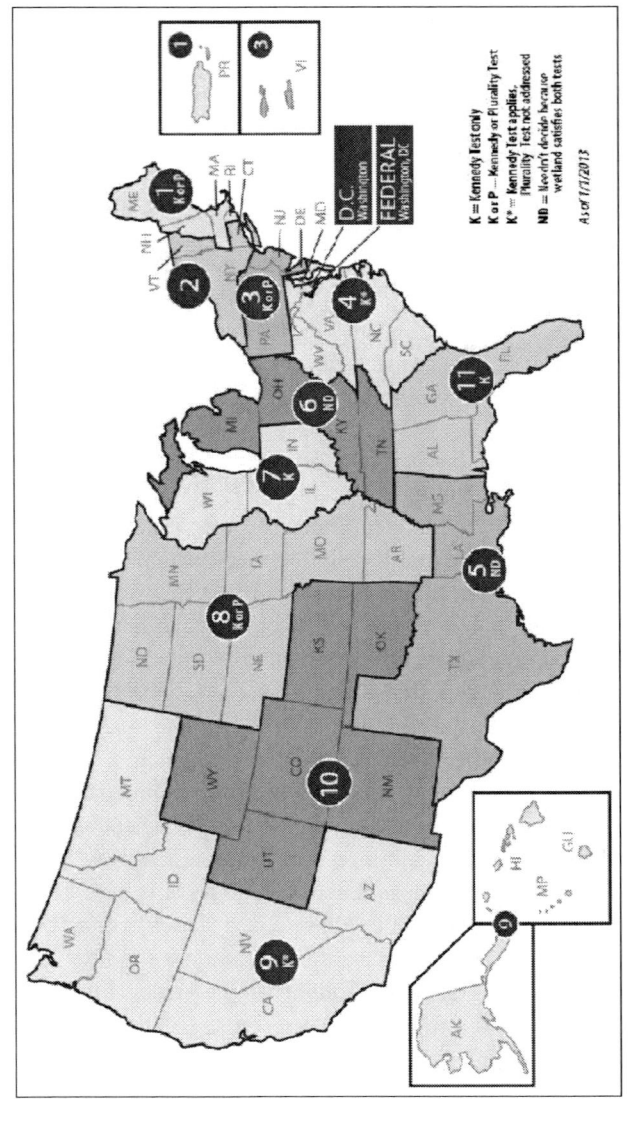

Source: Compiled by CRS
Notes: The 13 Federal Judicial Circuits (See 28 U.S.C.A. §28)

Figure A-1. Which *Rapanos* Test Governs? Rulings of Federal Circuit Courts of Appeal Only.

End Notes

[1] 547 U.S. 715 (2006).
[2] 33 U.S.C. §§1251-1387.
[3] See in particular two precursors of the CWA: Rivers and Harbors Act of 1899 §§10 (33 U.S.C. §403), and 13 (33 U.S.C. §407). Section 13 covers tributaries of navigable waters as well.
[4] William H. Rodgers, Jr., Handbook on Environmental Law 401 (1977) (footnotes omitted).
[5] P.L. 92-500. To be precise, the 1972 enactment was titled the Federal Water Pollution Control Act Amendments of 1972. It was only after the 1977 amendments thereto that the act as a whole became known as the Clean Water Act.
[6] CWA §502(7), 33 U.S.C. §1362(7).
[7] Conference report S.Rept. 92-1236 at 144, *reprinted in* 1972 U.S. Code Cong. & Admin. News 3776, 3822.
[8] 33 U.S.C. §1344.
[9] Section 301(a), 33 U.S.C. §1311(a), prohibits the discharge of any pollutant into navigable waters, except in compliance with various CWA sections, including Section 404.
[10] NRDC v. Callaway, 392 F. Supp. 685 (D.D.C. 1975).
[11] 40 Fed. Reg. 31320 (July 25, 1975), amending 33 C.F.R. part 209.
[12] 474 U.S. 121 (1985).
[13] 33 C.F.R. §328.3(a)(7). An identical EPA definition is at 40 C.F.R. §230.3(s)(7).
[14] 531 U.S. 159 (2001).
[15] 33 C.F.R. §328.3(a)(3). An identical EPA definition is at 40 C.F.R. §230.3(s)(3).
[16] 33 C.F.R. §328.3(a)(3).
[17] *Id.* (emphasis added).
[18] In *SWANCC* dictum, the Court stated: "In order to rule for the [Corps of Engineers], we would have to hold that the jurisdiction of the Corps extends to ponds that are *not* adjacent to open water. But we conclude that the text of the statute will not allow this." 531 U.S. at 168 (emphasis in original).
[19] The agencies did consider initiating a rulemaking to consider "issues associated with the scope of waters that are subject to the Clean Water Act" in light of *SWANCC*, 68 *Fed. Reg.* 1991 (2003), but the effort was abandoned in December 2003.
[20] *SWANCC*, 531 U.S. at 167.
[21] Corps of Engineers regulations define the word "adjacent" in "adjacent wetlands" to mean "bordering, contiguous, or neighboring. Wetlands separated from other waters of the United States by man-made dikes or barriers ... are 'adjacent wetlands.'" 33 C.F.R. §328.3(c).
[22] In addition to these three major opinions, Chief Justice Roberts wrote a brief opinion concurring with the plurality, and Justice Breyer wrote a brief opinion concurring with the dissenters.
[23] 547 U.S. at 722.
[24] *Id.* at 726.
[25] *Id.* at 732-733.
[26] *Id.* at 733-734.
[27] Hydrological connection is the test that the Corps has used to demonstrate significant nexus.
[28] Soon after *Rapanos* was decided, a federal district court commented that Justice Kennedy's opinion "advanced an ambiguous test—whether a 'significant nexus' exists to waters that are/were/might be navigable.... This test leaves no guidance on how to implement its vague, subjective centerpiece." United States v. Chevron Pipe Line Co., 437 F. Supp. 2d 605 (N.D. Tex. 2006).

[29] 547 U.S. at 780.

[30] *Id.*

[31] Going back to the CWA's enactment in 1972, several of the federal circuits have addressed issues as the scope of CWA jurisdiction—that is, the scope of "waters of the United States"—on ten or more occasions. See Marjorie A. Shields, *What Are "Navigable Waters Subject to Federal Water Pollution Control Act*," 160 A.L.R. Fed. 585 (updated weekly).

[32] United States v. Robison, 505 F.3d 1208 (11th Cir. 2007), *cert. denied*, 129 S. Ct. 627 (2008); United States v. Gerke Excavating, Inc., 464 F.3d 723 (7th Cir. 2006), *cert. denied*, 552 U.S. 810 (2007);

[33] Northern California River Watch v. Wilcox, 2011 Westlaw 238292, *1 (9th Cir. Jan. 26, 2011), *clarifying* Northern California River Watch v. City of Healdsburg, 496 F.3d 993 (9th Cir. 2007), *cert. denied*, 552 U.S. 1180 (2008); Precon Development Corp. v. U.S. Army Corps of Engineers, 2011 Westlaw 213052 (4th Cir. Jan. 25, 2011).

[34] United States v. Donovan, 661 F.3d 174 (3d Cir. 2011), *cert. denied*, 132 S. Ct. 2409 (2012); United States v. Bailey, 571 F.3d 791 (8th Cir. 2009); United States v. Johnson, 467 F.3d 56 (1st Cir. 2006), *cert. denied*, 552 U.S. 948 (2007).

[35] United States v. Cundiff, 555 F.3d 200 (6th Cir.), *cert. denied*, 130 S. Ct. 74 (2009); United States v. Lucas, 516 F.3d 316 (5th Cir.), *cert. denied*, 129 S. Ct. 116 (2008).

[36] See, e.g., United States v. Evans, 2006 Westlaw 2221629 (M.D. Fla. 2006) (Kennedy test or plurality test); Environmental Protection Information Center v. Pacific Lumber Co., 469 F. Supp. 2d 803 (S.D. Cal. 2007) (bound by *City of Healdsburg* to apply Kennedy test only); Simsbury-Avon Preservation Soc'y, LLC v. Metacon Gun Club, Inc., 472 F. Supp. 2d 219 (D. Conn. 2007) (Kennedy test or plurality test).

[37] One reported decision took its cue from the Scalia plurality view, though principally relying on circuit precedent. United States v. Chevron Pipe Line Co., 437 F. Supp. 2d 605 (N.D. Tex. 2006). This decision actually involved the amendments to the CWA made by the Oil Pollution Act, which uses the same definition of "waters of the United States" as CWA Section 404. A second decision holds that the significant nexus test is inapplicable outside the isolated wetlands context (with the implication that the plurality test alone applies). Sierra Club v. City and County of Honolulu, 2008 Westlaw 3850495, *7 (D. Hawaii August 18, 2008).

[38] Cruden, John C., Deputy Assistant Attorney General, Environment and Natural Resources Division, U.S. Department of Justice, "Statement Concerning Recent Supreme Court Decisions Dealing with the Clean Water Act," before the Subcommittee on Fisheries, Wildlife and Water, U.S. Senate Committee on Environment and Public Works, August 1, 2006, p. 16.

[39] See opinions of Justice Kennedy, Justice Breyer, and Chief Justice Roberts.

[40] U.S. Environmental Protection Agency, U.S. Army Corps of Engineers, "Clean Water Act Jurisdiction Following the U.S. Supreme Court's Decision in *Rapanos v. United States* & *Carabell v. United States*," December 2, 2008, p. 3.

[41] 547 U.S. at 738.

[42] "Clean Water Jurisdiction Following the Supreme Court's Decision in Rapanos v. United States and Carabell v. United States," Dec. 2, 2008, see http://water.epa.gov/lawsregs /guidance/wetlands/CWAwaters.cfm, under "Previous EPA Statements on Waters of the US." This webpage contains the 2008 guidance and the 2007 guidance, now superseded. It also includes a legal memorandum issued in January 2003 that continues to govern the agencies' interpretation of jurisdiction over the "isolated waters" addressed in the Supreme Court's 2001 *SWANCC* ruling.

[43] "Questions and Answers Regarding the Revised *Rapanos & Carabell* Guidance, December 2, 2008," p. 3, http://water.epa.gov/lawsregs/guidance/wetlands/upload/2008_12_5_wetlands_Rapanos_-20Guidance_QA20120208.pdf.

[44] These include all waters described in 33 C.F.R. §328.3(a)(1) (Corps of Engineers) and 40 C.F.R. §230.3(s)(1) (EPA). The 2008 guidance provides clarification of the scope of traditional navigable waters and guidance to field staff on making such a determination.

[45] The Corps has eight U.S. Divisions (which generally follow watershed boundaries), further subdivided into 38 Districts.

[46] U.S. General Accounting Office (now Government Accountability Office), "Waters and Wetlands, Corps of Engineers Needs to Evaluate Its District Office Practices in Determining Jurisdiction," February 2004, GAO-04-297.

[47] 547 U.S. at 775 (2006).

[48] American Rivers, "Bush Administration's So-called Revised Guidance on Clean Water is Just More of the Same," press release, December 3, 2008.

[49] Memorandum from Granta Nakayama, EPA Ass't Administrator for Enforcement and Compliance Assurance, to Benjamin Grumbles, EPA Ass't Administrator for Water, "OECA's Comments on the June 6, 2007 Memo, *Clean Water Act Jurisdiction Following the U.S. Supreme Court's Decision in Rapanos v. United States and Carabell v. United States*," dated March 4, 2008, on file with authors. EPA informs us that the "June 6, 2007 Memo" is the same as the June 2007 guidance referred to in footnote 42.

[50] Jeff Kinney, "Clean Water Act Jurisdictional Decisions Slower, More Complex, Amy Corps Says," *Daily Environment Report*, May 20, 2008, p. A-3.

[51] "Decline of Clean Water Act Enforcement Program," Majority Staff Memorandum to Representative Henry Waxman, Chairman, House Committee on Oversight and Government Reform, and Representative James L. Oberstar, Chairman, House Committee on Transportation and Infrastructure, December 16, 2008, 21 p., on file with authors.

[52] Environmental Protection Agency and Department of Defense, "EPA and Army Corps of Engineers Guidance Regarding Identification of Waters Protected by the Clean Water Act," 76 *Federal Register* 24479, May 2, 2011.

[53] See note 39 and accompanying text.

[54] Environmental Protection Agency and Army Corps of Engineers, "Draft Guidance on Identifying Waters Protected by the Clean Water Act," April 27, 2011, p. 2, on file with authors.

[55] *Id.* at 3.

[56] Stream order is a measure of the relative size of streams and is a method for stream classification based on relative position within a river network. Stream sizes range from the smallest, first-order, to the largest, the twelfth-order (e.g., the Amazon River). Headwater streams (first- and second-order) are the most abundant stream type in most river networks, and they supply most of the water in rivers. See http://www.cotf.edu/ete/modules/wqphysmethods.html.

[57] "Other waters" are waters that are geographically separated in that they do not have a hydrological connection to jurisdictional waters. Jurisdiction previously was asserted under Corps regulations (33 CFR §328.3(a)(3)) which provide for CWA jurisdiction over "[a]ll other waters ... the use, degradation or destruction of which could affect interstate or foreign commerce...." Geographically isolated, non-navigable intrastate waters were at issue in *SWANCC*.

[58] U.S. Environmental Protection Agency, "Potential Indirect Economic Impacts and Benefits Associated with Guidance Clarifying the Scope of Clean Water Act Jurisdiction," April 27, 2011, on file with authors.

[59] Inside E.P.A., "EPA's Draft Guidance Seeks To 'Increase Significantly' Water Act's Scope," February 17, 2011.

[60] Letter from Honorable Benjamin L. Cardin et al. to President Barack Obama, March 31, 2011, on file with authors.

[61] Letter from Honorable Bob Gibbs et al. to Lisa P. Jackson, EPA Administrator, and Jo-Ellen Darcy, Assistant Secretary of the Army for Civil Works, April 14, 2011; and Letter from Honorable John Barrasso et al. to Lisa P. Jackson, EPA Administrator, May 27, 2011, on file with authors.

[62] 16 U.S.C. §1536.

[63] "EPA Eyes Guide to Clarify Water Act's Scope for Discharge Permits," *Inside EPA*, Vol. 29, no. 10, March 7, 2008.

[64] Scott G. Leibowitz, "Isolated Wetlands and Their Functions: An Ecological Perspective," *Wetlands*, vol. 23, no. 3, September 2003, pp. 517-531.

[65] See, e.g., U.S. Congress, House of Representatives, Committee on Transportation and Infrastructure, Subcommittee on Water Resources and Environment, "Inconsistent Regulation of Wetlands and Other Waters," Hearing, 108th Congress, 2d Session, March 30, 2004 (H.Hrg. 108-58), 200 p.

[66] Jon Kusler, The Association of State Wetland Managers, "'Waters of the U.S.' After SWANCC," August 12, 2005 (draft), p. 6.

[67] Jon Kusler, The Association of State Wetland Managers, "The SWANCC Decision: State Regulation of Wetlands to Fill the Gap," March 2004, pp. 6-8. Hereafter, Kusler.

[68] Benjamin H. Grumbles, Assistant Administrator for Water, EPA, letter to Ms. Jeanne Christie, Association of State Wetland Managers, January 9, 2005 (sic), p. 3. The letter was written in January 2006, not 2005.

[69] American Rivers and Sierra Club, "Where Rivers Are Born: The Scientific Imperative for Defending Small Streams and Wetlands," February 2007, p. 7.

[70] Benjamin H. Grumbles, Assistant Administrator for Water, EPA, and John Paul Woodley, Assistant Secretary of the Army for Civil Works, Department of the Army, Statement before the Subcommittee on Fisheries, Wildlife, and Water of the U.S. Senate Committee on Environment and Public Works, August 1, 2006, 109th Congress, 2d session.

[71] James Murphy, "*Rapanos v. United States*: Wading Through Murky Waters," *National Wetlands Newsletter*, vol. 28, no. 5, September-October 2006, p. 19.

[72] U.S. Environmental Protection Agency, Office of Research and Development, *Connectivity of Streams and Wetlands to Downstream Waters: A Review and Synthesis of the Scientific Evidence*, External Review Draft, EPA/600/R-11- 098B, September 2013, http://yosemite.epa.gov/sab/sabproduct.nsf/0/7724357376745F48852579E60043E88C/$File / WOUS_ERD2_Sep2013.pdf.

[73] See U.S. Environmental Protection Agency, "Clean Water Act Definition of 'Waters of the United States,'" http://water

[74] Amena H. Saiyid, "EPA Study Could Be Used to Expand Reach of Law Over Waters, Wetlands," *Daily Environment Report*, September 27, 2013, pp. BB-1.

[75] U.S. Environmental Protection Agency, "Advance Notice of Proposed Rulemaking on the Clean Water Act Regulatory Definition of 'Waters of the United States,'" 68 *Federal Register* 1994-95, January 15, 2003.

[76] Kusler, p. 15.

[77] Jan Goldman-Carter, "Isolated Wetland Legislation: Running the Rapids at the State Capitol," *National Wetlands Newsletter*, May-June 2005, pp. 27-29.

[78] Turner Odell, "On Soggy Ground—State Protection for Isolated Wetlands," *National Wetlands Newsletter,* September-October 2003, p. 10.

[79] See http://epw.senate.gov/public/index.cfm?FuseAction=Majority.PressReleases&ContentRecord_id=64739ae3- 802a-23ad-4c30-36fc58cc1014&Region_id=&Issue_id=.

[80] The committee report on the bill, S.Rept. 111-361, was filed in December 2010, 18 months after the committee's action to approve the amended legislation. Companion legislation was introduced in the House in the 111th Congress (H.R. 5088), but no further action occurred.

In: U.S. Wetlands
Editor: Harriet M. Hutson

ISBN: 978-1-63117-800-9
© 2014 Nova Science Publishers, Inc.

Chapter 3

THE SUPREME COURT ADDRESSES CORPS OF ENGINEERS JURISDICTION OVER "ISOLATED WATERS": THE SWANCC DECISION[*]

Robert Meltz and Claudia Copeland

SUMMARY

On January 9, 2001, the Supreme Court handed down *Solid Waste Agency of Northern Cook County (SWANCC) v. U.S. Army Corps of Engineers*. At issue in *SWANCC* was the scope of Clean Water Act section 404, which requires permits for the discharge of dredged or fill materials into "navigable waters," defined by the Act as "waters of the United States." Section 404 is the charter for the federal wetlands permitting program.

SWANCC explicitly held that the Corps of Engineers' use of the "migratory bird rule," adopted by the agency to interpret the reach of its section 404 authority over "isolated waters" (including isolated wetlands), exceeded the authority granted by that section. Looking at the decision's rationale rather than its holding, however, *SWANCC* may be read more broadly to bar assertion of section 404 jurisdiction over isolated waters on *any* basis, migratory bird rule or otherwise. The

---

[*] This is an edited, reformatted and augmented version of a Congressional Research Service publication, CRS Report for Congress RL30849, dated February 16, 2001.

migratory bird rule asserted that section 404 covers, among other waterbodies, isolated waters "which are or would be used as habitat by ... migratory birds that cross state lines ...."

In 1985, the Supreme Court had sustained the assertion by the Corps and EPA that waters and wetlands *adjacent to* navigable waters, interstate waters, or their tributaries are "waters of the United States" under section 404. The question left for *SWANCC* was whether waters and wetlands not so adjacent – "isolated waters" – also are so covered. Such jurisdictional lines stand in contrast to the scientists' perspective, which recognizes the value of wetlands based on water quality and other physical functions which they perform, irrespective of whether the wetlands are isolated or contiguous to other waters.

Estimates of waters and wetland acreage likely to be removed from the section 404 permitting program as a result of the *SWANCC* decision are very difficult to assess, in part because of questions about Corps and EPA interpretation of the ruling, but the decision may affect up to 79% of wetland acreage. One likely result is that in those cases where case-by-case evaluations will be required to determine if regulatory jurisdiction exists, the length of time to obtain section 404 permits will be longer than in the past. If federal jurisdiction is diminished, the responsibility to protect affected wetlands falls on states and local governments. A comprehensive picture of their ability to protect wetlands, under various possible state and local authorities, is difficult to draw together. Whether states will act to fill in the gap left by removal of some federal jurisdiction through new laws or programs raises difficult political and resource questions.

The *SWANCC* decision also raises issues for Congress. First is whether confusion that may now exist about the extent of Clean Water Act jurisdictional waters and wetlands should be resolved, and what constitutional limits may apply. Second is whether to provide federal resources and incentives to encourage expansion of state wetlands protection and regulatory programs or others that encourage acquisition and conservation of wetlands.

INTRODUCTION

On January 9, 2001, the Supreme Court added another to its growing list of recent decisions on the proper line of demarcation between federal and state authority in our dual-sovereign system of government. In *Solid Waste Agency of Northern Cook County (SWANCC) v. U.S. Army Corps of Engineers*,[1] the Court addressed the geographic scope of Clean Water Act (CWA) section 404,[2] which requires permits for the discharge of dredged or fill materials into "navigable waters," defined by the Act as "waters of the United States," and is the charter for the federal wetlands permitting program.

SWANCC expressly held that the Corps of Engineers' use of the long controversial "migratory bird rule," adopted by the Corps and Environmental Protection Agency (EPA) in 1986 to interpret the reach of their authority over discharges into "isolated waters," exceeded the authority granted by section 404. As we will discuss, however, the decision's rationale may easily be read more broadly to bar assertion of 404 jurisdiction over isolated waters on *any* theory, migratory bird rule or otherwise.

This report discusses the background of the case, the Supreme Court's decision, and the legal and policy implications of the ruling for the CWA, especially for section 404. It discusses the ecological services and physical functions performed by wetlands in maintaining water quality, including in isolated waters. States' ability to fill in any regulatory gaps resulting from the decision and issues for Congress also are addressed.

BACKGROUND OF THE CASE

SWANCC, a consortium of Chicago-area cities and villages, sought to develop a landfill for baled nonhazardous solid waste on a 533-acre parcel in Illinois. The parcel had been used for sand and gravel mining until about 1960. Since then, the excavation trenches from the mining had evolved into ponds ranging in size from a few feet across to several acres. SWANCC obtained the needed local and state permits, but the Corps, based on the ponds and their use by migratory birds, asserted jurisdiction under section 404 and denied a permit.

Section 404 requires permits for discharges to dispose of dredged and fill material into the nation's navigable waters, such as when a landowner undertakes activity to develop or otherwise improve his or her property. To assess whether this requirement applies to a particular activity, a landowner must determine whether the disposal site is a "water of the United States" within CWA jurisdiction. The definitions of waters subject to CWA jurisdiction are contained in regulations of the Corps of Engineers and EPA, the agencies with primary responsibility for administering section 404.[3] Through judicial interpretation and regulatory changes since the 1970s, the types of regulated waters have evolved from narrow to broad, and also to include wetlands. Congress has not amended section 404 since 1977, when it provided regulatory exemptions for categories of routine activities, such as normal farming and forestry.

The SWANCC site ponds are known in section 404 parlance as "isolated waters"– waters that are not traditionally navigable or interstate, nor tributaries thereof, nor adjacent to any of these. Long ago, the Supreme Court in *United States v. Riverside Bayview Homes, Inc.* upheld the Corps' authority under section 404 to regulate wetlands (and other waters) *adjacent to* navigable and interstate waters, and their tributaries.[4] It expressly left open the question, however, whether isolated waters, not being adjacent, lie within the reach of section 404, or, for that matter, within Congress' power under the Commerce Clause of the Constitution.[5] Both before and after *Riverside Bayview*, the lower courts have wrestled with these questions.

The Corps' assertion of jurisdiction over the isolated waters at the SWANCC site, as elsewhere, was based on a three-step argument. First, section 404 applies by its terms to "navigable waters," defined expansively by the CWA to mean "the waters of the United States."[6] Second, under 1977 regulations the Corps defines "waters of the United States" broadly to include, in addition to traditionally navigable waters, interstate waters, their tributaries, and adjacent wetlands, the following –

> [all] other waters such as intrastate lakes, rivers, streams (including intermittent streams), mudflats, sandflats, wetlands, sloughs, prairie potholes, wet meadows, playa lakes, or natural ponds, the use, degradation, or destruction of which could affect interstate commerce ....[7]

Third, the Corps' migratory bird "rule," a 1986 attempt by the agency to clarify the intrastate waters covered by this regulation, says that such "isolated waters" include those "which are or would be used as habitat by ... migratory birds that cross state lines ...."[8] The Corps had found that the water areas on the SWANCC site are used as habitat by migratory birds that cross state lines.

In reading its section 404 jurisdiction broadly, the Corps was not without congressional support. In defining "navigable waters" as "waters of the United States," Congress "evidently intended to repudiate limits that had been placed on federal regulation by earlier water pollution control federal statutes ...."[9] Indeed, the conference report accompanying enactment of the CWA in 1972 states that "[t]he conferees fully intend that the term 'navigable waters' be given the broadest possible constitutional interpretation ...."[10]

The district court in *SWANCC* granted summary judgment to the Corps of Engineers on the jurisdictional issue. On appeal, the U.S. Court of Appeals for the Seventh Circuit ruled in favor of Corps jurisdiction as well. The Seventh Circuit found that Congress has the authority under the Commerce Clause of the Constitution[11] to regulate isolated waters, and that Congress, in enacting section 404, intended to reach such waters. The Supreme Court reversed.

THE SUPREME COURT DECISION

As with many other Supreme Court decisions involving the line between federal and state power, the *SWANCC* ruling saw the Court divide into now-familiar 5-4 voting blocs. The five-justice majority opinion, in one reading, concluded only that the Corps and EPA could not continue to use the migratory bird rule to assert section 404 jurisdiction over isolated waters. "We conclude," said the Court at one point, that the `Migratory Bird Rule' is not fairly supported by the CWA."[12] The decision's rationale, however, was broader, appearing to preclude federal assertion of section 404 jurisdiction over isolated waters on *any* basis. Stated the Court: "In order to rule for [the Corps], we would have to hold that the jurisdiction of the Corps extends to ponds that are *not* adjacent to open water. But we conclude that the text of the statute will not allow this."[13]

In any event, the Court deemed it unnecessary to reach the constitutional issue pressed by SWANCC: whether, had the migratory rule been authorized under the CWA, it exceeded Congress' power under the Commerce Clause.

The majority opinion, written by Chief Justice Rehnquist, held that Congress, in enacting the 1977 amendments to the CWA, had not implicitly approved the Corps' broad definition of "navigable water" adopted that year[14] under the original 1972- enacted CWA. For example, Congress' failure to pass a bill in 1977 containing a narrow definition of navigable waters had not been shown by the Corps, said the majority, to constitute congressional approval of the Corps' broad definition. The majority then declined to afford the Corps the customary deference granted agency interpretations of ambiguous statutes. For one thing, it said, section 404 is not ambiguous at all. And even if it were, deference is not appropriate where an agency interpretation of a statute "invokes the outer limits of Congress' power" – a reference to the Court's milestone decisions in recent years involving the reach of the Commerce Clause. This concern is particularly strong, it said, where the agency interpretation permits encroachment on a traditional state power – here, that over land and water use.

The dissent, penned by Justice Stevens, asserts that given *Riverside Bayview*'s holding that the CWA went beyond navigable waters, embracing marshes and inland lakes adjacent thereto, there is no principled reason to stop there. The 1972 CWA, in the dissent's view, offers no support for such a constraint, and the 1977 CWA amendment supports coverage of isolated waters. Moreover, the dissent declares, there is no Commerce Clause problem, since the discharge of dredged and fill materials into "waters of the United States" constitutes an economic activity that may be aggregated to show a substantial effect on interstate commerce.

LEGAL IMPLICATIONS OF THE DECISION

The *SWANCC* decision continues the efforts of the five Supreme Court justices generally regarded as conservative to limit federal regulatory power. In 1995 and 2000, these same five justices found that Congress had exceeded Commerce Clause limits in enacting legislation dealing with possession of guns in school zones[15] and violence against women.[16] To be sure, the Supreme Court in *SWANCC* did not reach the constitutional question, but rather disposed of the case on purely statutory grounds. Nonetheless, its analysis of the CWA has, as noted, a strong undercurrent of the same Commerce Clause and federalism concerns. In not addressing the constitutional issue, *SWANCC* resembles another recent Commerce Clause decision where, as in *SWANCC*, the Court used the possibility of Commerce Clause issues being raised by a broad interpretation of a federal statute to support its adoption of a narrow reading.[17] The next event to watch for in this constitutional area is the Court's decision whether to grant the pending petition for certiorari in a Commerce Clause challenge to the Endangered Species Act.[18]

The problem underlying the *SWANCC* decision arises largely from the history of some jurisdictional terms used by Congress in water-related statutes. In the nineteenth century and first half of the twentieth century, Congress set the scope of many such statutes as the "navigable waters" of the United States. This made sense because Congress was focussed on fostering waterborne commerce, and the Supreme Court had obliged by articulating a broad vision of federal power over navigation. A prominent example, and a forerunner of the CWA, was the Rivers and Harbors Act of 1899, particularly its section 13 known as the "Refuse Act."[19] With the shift in emphasis in the mid-twentieth century from protection of navigation to protection of the environment, however, the phrase "navigable waters" was no longer a comfortable fit.

Notwithstanding, Congress used "navigable waters" in 1972 when it wrote the CWA, including section 404, accommodating the broader concerns of environmental protection by defining "navigable waters" expansively to mean "waters of the United States." The question in *SWANCC* was whether this definition entirely removed the "navigable" qualifier from the Act, or merely limited it. The majority justices opted for the latter, and were unwilling to go beyond the erosion of this qualifier already accepted by the Court in *Riverside Bayview*.

SWANCC's implications for the scope of the federal wetlands permitting program are certain to be significant, but it will take years of litigation before they are fully clear. A key source of confusion is the aforementioned disconnect between the decision's narrow holding and broad rationale. The latter appears to preclude any effort by the Corps to assert jurisdiction over isolated waters, including isolated wetlands, based on linkages with interstate commerce other than the interstate flight of migratory birds. (The Corps did assert such non-migratory-bird linkages after the litigation commenced, noting that SWANCC's municipal landfill is clearly of a commercial nature and, when aggregated with similar activity elsewhere, would substantially affect interstate commerce. The Supreme Court declined to consider this argument, pointing out that landfill activity is a "far cry" from the "waters of the United States" to which the CWA extends.) Plainly, the degree of section 404 program contraction occasioned by *SWANCC* will depend on which aspects of the decision shape the government's response.

On that point, a legal memorandum issued by the Corps and EPA on January 19, 2001,[20] hews more closely to the narrow holding, though in tentative terms. The memorandum notes the above-discussed difference between the *SWANCC* rationale and holding, and, presumably as a result, takes a case-by-case approach as to the waters falling within the coverage gap (last item below). In summary, the memorandum asserts that –

- Traditionally navigable waters, interstate waters, their tributaries, and wetlands adjacent to each are still covered.
- Intrastate waters that could affect interstate commerce *solely* by virtue of their use as habitat for migratory birds are no longer covered.
- As to intrastate waters having other (non-migratory bird) connections to interstate commerce, staff is advised to consult agency legal counsel. For example, waters that are isolated and intrastate, but nonetheless navigable (such as the Great Salt Lake in Utah) may still

support jurisdiction "if their use, degradation, or destruction could affect interstate or foreign commerce ...."

The guidance goes on to state its view that the *SWANCC* holding, while important, is "limited," and must be interpreted in light of other Supreme Court precedents "which ... broadly uphold CWA jurisdictional authority."

Another source of uncertainty for the wetlands permitting program will be the extent to which, lacking any clear authority over isolated waters now, the Corps may seek to recharacterize wetlands from that category to ones over which its section 404 authority remains undisputed. For example, the concept of traditionally navigable waters is an elastic one, covering all waters that are now navigable, were once navigable, or could reasonably be made navigable in the future.[21] Another example is the "adjacent wetlands" jurisdiction upheld in *Riverside Bayview*. One might think the concept of adjacency to be relatively clear, but in *Riverside Bayview* itself, the wetlands in question were only "near" the shores of the lake – "*part of* a wetland that actually abuts on a navigable waterway."[22] The Court expressly noted that the concept of adjacent wetlands includes those adjacent wetlands "that are not the result of flooding or permeation by water having its source in adjacent bodies of water,"[23] approving the Corps view that wetlands may affect the water quality of adjacent waterbodies by functioning as integral parts of the same aquatic environment.

It is not known whether the newly arrived Bush Administration would support such recharacterization efforts.

As a final legal matter, we note that by interpreting the scope of "waters of the United States" as used in section 404, *SWANCC* will affect the scope of other CWA sections whose jurisdictional scope is defined by that same phrase. Such sections include those governing oil spill cleanup (section 311), the National Pollutant Discharge Elimination System permit program (section 402), and state water quality certification (section 401).

CWA JURISDICTION:
WATERS AND ISOLATED WETLANDS

Long before the *SWANCC* case, there had been controversy and litigation over whether isolated waters that are not adjacent to true navigable waters are properly within the jurisdiction of section 404. Prior to *SWANCC*, virtually all

U.S. wetlands were, at least theoretically, subject to regulation under section 404. Philosophically, the debate about isolated waters concerns whether the waters themselves and wetlands not adjoining surface waters contribute to the CWA's goals of restoring and maintaining the chemical, physical, and biological integrity of the nation's waters. In the case of isolated waters that are wetlands, these questions are exacerbated because wetlands may be seasonally dry or reduced in area and may not have the physical characteristics of wetlands at all times. Some wetlands (swamps, bogs, and marshes) are easily identified by the public as wetlands. Others are less obvious and require sophisticated technical evaluation to determine that they have the three factors (water, hydric or wet soil, and vegetation adapted for or requiring wetland conditions) that scientists generally agree define wetlands.[24] The result, in cases where wetlands are less clearly identifiable, can be confusion and conflict between landowners and regulators over national environmental goals and economic and property ownership goals and values. Yet even these less obvious wetlands perform valuable ecological services.

Scientists recognize the value of wetlands based on a range of physical functions that they perform. One group of functions relates to water quality. Wetlands are good water filters: they remove and retain nutrients; they process chemical and organic wastes; and they reduce sediment loads to receiving waters. Wetlands also provide flood damage protection to urban and agricultural lands by storing flood waters that overflow river banks or surface waters and by collecting waters in isolated depressions. Wetlands recharge groundwater reserves that are hydrologically connected to surface waters. According to a 1995 National Research Council report, many of these functions occur irrespective of whether the wetlands are isolated or contiguous to navigable waters, because of groundwater connections between isolated wetlands and surface waters. Small, shallow wetlands that are isolated from rivers are frequently important to waterfowl, the NRC said, for food and forage. Also, sites that are intermittently flooded (even those that may be completely dry for several years) can be important for storing flood waters and can have distinctive water-dependent biota (plants and animals) that persist over dry intervals but return when water returns to the site.[25]

POLICY IMPLICATIONS OF THE *SWANCC* RULING

Since the Court's actual holding concerning CWA regulation was narrow, while its rationale was wider ranging (as discussed above), the policy implications of how much the *SWANCC* decision restricts federal regulation depend on how broadly or narrowly the opinion is applied. Two scenarios are possible. A broad reading would be interpreted as knocking out all section 404 jurisdiction and Corps regulation of isolated waters and wetlands. But a narrow reading, one asserting that jurisdiction will be found lacking only if the sole connection to interstate commerce is the presence of migratory birds and wildlife, would allow federal regulation of some isolated waters to continue (e.g., in waters that are used by interstate travelers for recreation).

The question of which view the government would take was answered in a January 19 memorandum issued jointly by EPA and the Corps for headquarters and field/regional staff who work on the section 404 program, discussed above, in which the agencies provide a legal interpretation of *SWANCC* based on a narrow reading of the Court's decision. Some observers anticipate that the agencies' interpretation will be tested in the courts.

A key policy question that may not be clearly answered for some time is how regulatory protection of wetlands will be affected or reduced as a result of the decision. Many types of isolated wetlands are not physically adjacent to navigable waters and under a broad reading of the decision, would lack regulation. Major wetland types that potentially would not be regulated include prairie potholes of the Upper Midwest, wet meadows, river fringing wetlands along small nonnavigable rivers and streams, lake fringing wetlands for smaller nonnavigable lakes, many forested wetlands, playas and vernal ponds of Texas and other areas of the west, seeps and spring, flats, bogs and large amounts of tundra in Alaska. A new report by the Department of Interior's Fish and Wildlife Service estimates that in 1997, there were 105.5 million acres of wetlands on public and private lands in the conterminous United States and that between 1986 and 1997, a net of 644,000 acres of wetlands was lost.[26] According to an analysis prepared by the Association of State Wetland Managers (ASWM), accurate estimates of impacts of the decision on wetland resources are not possible, in part because of uncertainty about how key terms in the opinion (such as "adjacent" and "tributary") will be defined, whether broadly or narrowly. Still, ASWM believes that impacts are likely to be environmentally significant.

> Tentative state estimates which have been provided to the Association of State Wetland Managers suggest 30% to 79% of total wetland acreage may be affected...Even if SWANCC results in only a one percent loss of America's wetlands, the decision would cause more wetlands to be destroyed than were lost in the past decade.[27]

While others may believe that a lesser percentage of the nation's wetlands would be so affected, there are no government estimates for comparison. In terms of regulatory processes, however, one likely result of the *SWANCC* decision is that in those cases where case-by-case evaluations will be required to determine whether the use, degradation or destruction of the waters in question could affect interstate commerce and, thus, if jurisdiction exists, the length of time for landowners to obtain section 404 permits will be longer than in the past.

The *SWANCC* decision affects not only privately owned lands but also isolated waters and wetlands located on public lands: the federal government owns about one-third of the nation's lands. As a result of the Court's decision, federal agency decisions on these lands affecting isolated wetlands will no longer be subject to section 404 permitting, although they will be subject to requirements of the National Environmental Policy Act and Executive Orders dealing with wetlands, floodplain management, and protection of migratory birds.

In addition to the section 404 program, questions arise about impacts of the *SWANCC* ruling on other parts of the CWA, especially its principal permit program, the National Pollutant Discharge Elimination System (NPDES) program under section 402. It requires permits for pollutant discharges from point sources (industrial facilities and municipal sewage plants) into the nation's waters. Another provision, section 311, concerns liability for oil discharges into the nation's waters. Neither of these was at issue in this case but might be challenged through extension of the ruling. The January 19 joint Corps-EPA memorandum states that federal implementation of any other CWA provision that involves "waters of the United States" will be governed by the same interpretation that applies to section 404. As a result, federal jurisdiction to require NPDES permits or assess oil spill liability in some isolated waters could be limited. These impacts will become clearer in time.

STATE AUTHORITY

As noted previously, prior to *SWANCC*, virtually all U.S. wetlands were, at least theoretically, subject to regulation under section 404. As federal jurisdiction is diminished, the responsibility to protect affected wetlands falls on states primarily and local governments, which also regulate some wetlands. State and local wetlands regulatory programs focus primarily upon navigable waters, tributaries, and adjacent wetlands. They supplement but do not substitute for federal jurisdiction. According to the Association of State Wetland Managers (ASWM),[28] 14 states have some form of regulatory program for freshwater wetlands, but they are quite variable. Differences exist in part because freshwater wetland types vary greatly across the nation and because of differing state preferences. Some of the state programs are very comprehensive, but regulations in many of the 14 states are limited by wetland size, mapping requirements, and exemptions for specified activities. According to ASWM, state regulations do not generally apply to federal lands. Some of the states with the largest isolated wetland acreages provide little or no state protection, including Alaska, Louisiana, Texas, North Dakota, South Dakota, South Carolina, North Carolina, Georgia, Nebraska, Kansas, and Mississippi.

Changes in section 404 jurisdiction would diminish use of one tool used by many states to control activities affecting wetlands. In recent years, most states have utilized CWA section 401 water quality certification programs in addition to or in lieu of specific regulatory statutes. Section 401 requires that, before a federal permit or license is issued, states must certify that the project complies with water quality standards. This authority effectively gives states a veto power on the federal permit or the ability to require conditions that become part of a permit. State water quality certification has been used by a number of states to control activities affecting wetlands without having to independently establish state permitting and enforcement programs. But, where federal jurisdiction does not exist and no section 404 or other federal permit is required, section 401 certification also is not required and thus is not available as a tool for the state to evaluate the proposed activity.

A number of states that do not have wetlands laws on their books do have other state environmental laws dealing with water quality or natural resources, and these may already provide substantial authority to regulate wetlands.[29] However, a comprehensive picture of states' ability to protect wetlands, under various possible authorities, is difficult to draw together. To fill in the gap left by removal of some federal regulatory jurisdiction, states could adopt more comprehensive wetlands regulatory statutes or wetland amendments to state pollution control statutes (possibly including independent water quality certification programs) and rules. The latter, for example, could integrate wetlands, water quality, and watershed management. States do not have the constitutional constraints that the federal government does in enacting legislation (i.e., whether a legislative action exceeds Congress' power under the Commerce Clause). Also, CWA section 404(t) expressly provides that the existence of section 404 does not preempt state law governing the discharge of dredged or fill material. However, whether states will take steps to expand wetlands protection in response to the Court's decision raises difficult political and resource questions (e.g., budget and staffing). It is quite likely that, among states, the *SWANCC* decision pleases some and is opposed by others.

ISSUES FOR CONGRESS

The Court's ruling raises issues for Congress. First is whether confusion that may now exist about the extent of CWA jurisdictional waters should be resolved. The Court said that Congress has not authorized the Corps to regulate isolated waters that are not adjacent to navigable waters. Some policymakers who hold the view that Congress intended that the law be interpreted broadly may want to amend it to make clear that section 404 and the CWA generally do apply to isolated wetlands and waters. Whether such an action would be considered constitutional by the Supreme Court remains to be seen. Others who favor an interpretation that limits federal regulatory authority may find fault with EPA's and the Corps' reading of the *SWANCC* decision and may want to amend the law to ensure a narrow jurisdictional interpretation.

Second is the issue of providing resources and incentives to encourage expansion of state wetlands protection and other programs. Other than Michigan and New Jersey, states have been reluctant to seek delegation of section 404 program authority to them both because of the resource burden that would be required and because the law currently does not allow states to issue permits for activities affecting traditionally navigable waters. To the extent the *SWANCC* decision limits the scope of the federal program, states are likely to be even less interested in assuming responsibility to issue section 404 permits.[30] Some may want Congress to address how to assist states in protecting wetlands through financial support (such as the EPA state wetlands grant program, currently funded at $15 million per year), technical assistance, or other possible incentives. Congress also could expand existing programs, such as the Wetland Reserve Program, that provide incentives to private landowners for protection of isolated wetlands through acquisition and easements, rather than regulation.

Finally, as noted above, the views of the new Bush Administration on these issues are unknown for now. The Supreme Court's *SWANCC* decision and the Corps' and EPA's interpretation of it preceded Inauguration Day. Thus, it is unclear what directives and guidance, if any, the White House will provide to the agencies or how the Administration might respond to legislative initiatives in Congress or other possible proposals concerning state wetland or incentive programs.

End Notes

[1] 121 S. Ct. 675 (2001).

[2] 33 U.S.C. § 1344.

[3] The Corps administers the permit program under section 404, pursuant to EPA guidelines. CWA § 404(b); 33 U.S.C. § 1344(b). EPA also has veto authority over Corps permitting decisions, though it is rarely exercised. CWA § 404(c); 33 U.S.C. § 1344(c).

[4] 474 U.S. 121 (1985).

[5] Id. at 131 n.8.

[6] CWA § 502(7); 33 U.S.C. § 1362(7).

[7] 33 C.F.R. § 328.3(a)(3).

[8] 51 Fed. Reg. 41,206, 41,217 (1986) (in preamble).

[9] Riverside Bayview, 474 U.S. at 133.

[10] S. Rep. No. 92-1236, at 144 (1972).

[11] U.S. Const. art. I, sec. 8, cl. 3: "The Congress shall have Power ... To regulate Commerce ... among the several States ...." Because the Constitution nowhere confers an express authority on Congress to legislate for environmental protection, most federal environmental statutes rest on the broad, contemporary reading of the Commerce Clause.

[12] 121 S. Ct. at 680. See also 121 S. Ct. at 684 ("We hold that 33 CFR §328.3(a)(3) (1999), as clarified and applied to petitioner's balefill site pursuant to the 'Migratory Bird Rule,' 51 Fed. Reg. 41217 (1986), exceeds the authority granted to respondents under §404(a) of the CWA.").

[13] 121 S. Ct. at 680 (emphasis in original).

[14] See text accompanying note 7.

[15] United States v. Lopez, 514 U.S. 549 (1995).

[16] United States v. Morrison, 120 S. Ct. 1740 (2000).

[17] Jones v. United States, 120 S. Ct. 1904 (2000).

[18] Gibbs v. Babbitt, 214 F.3d 483 (4th Cir. 2000), petition for cert. filed, 69 U.S.L.W. 3383 (Nov. 22, 2000) (No. 00-844).

[19] 33 U.S.C. § 407.

[20] Memorandum by Gary Guzy, General Counsel, EPA, and Robert W. Andersen, Chief Counsel, Corps of Engineers, "Supreme Court Ruling Concerning CWA Jurisdiction over Isolated Waters" (Jan. 19, 2001). 9 p.

[21] United States v. Appalachian Power Co., 311 U.S. 377, 408 (1940).

[22] 474 U.S. at 124, 135 (emphasis added).

[23] Id. at 134.

[24] National Research Council. Wetlands: Characteristics and Boundaries. National Academy Press, 1995: 50-55, 79-131.

[25] Id. at 137-139.

[26] U.S. Fish and Wildlife Service. "Report to Congress on the Status and Trends of Wetlands in the Conterminous United States 1986 to 1997." January 2001, 84 p.

[27] Kusler, Jon, Associate Director, Association of State Wetlands Managers. "The SWANCC Decision and State Regulation of Wetlands." Memorandum, Feb. 7, 2001: 8-9.

[28] Id. at 9-10.

[29] California, for example, does not have a wetlands law and has used section 401 certification to evaluate projects affecting wetlands. However, according to a memorandum prepared by the chief counsel of the State Water Resources Control Board, irrespective of how the SWANCC decision is interpreted, the state retains independent authority under its statutes to regulate discharges of waste to all waters of the state, including those waters that are no longer considered waters of the United States. Craig M. Wilson, chief counsel, California State Water Resources Control Board. "Effect of SWANCC v. United States on the 401 Certification Program." Memorandum, Jan. 25, 2001. 5 p.

[30] CWA section 404(g) currently authorizes EPA, in consultation with the Corps, to delegate the federal section 404 permitting function to qualified states. Only New Jersey and Michigan have, so far, sought and received this authority. Other states cite a number of factors for not seeking program authorization, including the resource burden required to staff and operate a separate state office. Further, section 404(g)(1) provides that state authorization may not include traditionally navigable waters, so even before SWANCC, states' authority to administer the federal 404 permit program has been limited and is more restricted by the diminished jurisdiction resulting from the Court's decision.

In: U.S. Wetlands
Editor: Harriet M. Hutson

ISBN: 978-1-63117-800-9
© 2014 Nova Science Publishers, Inc.

Chapter 4

CLEAN WATER ACT: A SUMMARY OF THE LAW[*]

Claudia Copeland

SUMMARY

The principal law governing pollution of the nation's surface waters is the Federal Water Pollution Control Act, or Clean Water Act. Originally enacted in 1948, it was totally revised by amendments in 1972 that gave the act its current dimensions. The 1972 legislation spelled out ambitious programs for water quality improvement that have since been expanded and are still being implemented by industries and municipalities.

This report presents a summary of the law, describing the statute without discussing its implementation.

The Clean Water Act consists of two major parts, one being the provisions which authorize federal financial assistance for municipal sewage treatment plant construction. The other is the regulatory requirements that apply to industrial and municipal dischargers. The act has been termed a technology-forcing statute because of the rigorous demands placed on those who are regulated by it to achieve higher and higher levels of pollution abatement under deadlines specified in the law. Early on, emphasis was on controlling discharges of conventional

---

[*] This is an edited, reformatted and augmented version of a Congressional Research Service publication, CRS Report for Congress RL30030, from www.crs.gov, prepared for Members and Committees of Congress, dated November 30, 2012.

pollutants (e.g., suspended solids or bacteria that are biodegradable and occur naturally in the aquatic environment), while control of toxic pollutant discharges has been a key focus of water quality programs more recently.

Prior to 1987, programs were primarily directed at point source pollution, that is, wastes discharged from discrete sources such as pipes and outfalls. Amendments to the law in that year authorized measures to address nonpoint source pollution (runoff from farm lands, forests, construction sites, and urban areas), which is estimated to represent more than 50% of the nation's remaining water pollution problems. The act also prohibits discharge of oil and hazardous substances into U.S. waters.

Under this act, federal jurisdiction is broad, particularly regarding establishment of national standards or effluent limitations. Certain responsibilities are delegated to the states, and the act embodies a philosophy of federal-state partnership in which the federal government sets the agenda and standards for pollution abatement, while states carry out day-to-day activities of implementation and enforcement.

To achieve its objectives, the act is based on the concept that all discharges into the nation's waters are unlawful, unless specifically authorized by a permit, which is the act's principal enforcement tool. The law has civil, criminal, and administrative enforcement provisions and also permits citizen suit enforcement.

Financial assistance for constructing municipal sewage treatment plants and certain other types of water quality improvements projects is authorized under title VI. It authorizes grants to capitalize State Water Pollution Control Revolving Funds, or loan programs. States contribute matching funds, and under the revolving loan fund concept, monies used for wastewater treatment construction are repaid to states, to be available for future construction in other communities.

INTRODUCTION

The principal law governing pollution of the nation's surface waters is the Federal Water Pollution Control Act, or Clean Water Act. Originally enacted in 1948, it was totally revised by amendments in 1972 that gave the act its current shape. The 1972 legislation spelled out ambitious programs for water quality improvement that have since been expanded and are still being implemented by industries, municipalities and others. Congress made fine-tuning amendments in 1977, revised portions of the law in 1981, and enacted further amendments in 1987.

This report presents a summary of the law, describing the statute. Many details and secondary provisions are omitted here, and even some major components are only briefly mentioned. Further, this report describes the statute, while other CRS products discuss implementation issues.[1] *Table 1* shows the original enactment and subsequent major amendments. *Table 2*, at the end of this report, cites the major U.S. Code sections of the codified statute.

Table 1. Clean Water Act and Major Amendments (codified generally as 33 U.S.C. §§1251-1387)

Year	Act	Public Law
1948	Federal Water Pollution Control Act	P.L. 80-845 (Act of June 30, 1948)
1956	Water Pollution Control Act of 1956	P.L. 84-660 (Act of July 9, 1956)
1961	Federal Water Pollution Control Act Amendments	P.L. 87-88
1965	Water Quality Act of 1965	P.L. 89-234
1966	Clean Water Restoration Act	P.L. 89-753
1970	Water Quality Improvement Act of 1970	P.L. 91-224, Part I
1972	Federal Water Pollution Control Act Amendments	P.L. 92-500
1977	Clean Water Act of 1977	P.L. 95-217
1981	Municipal Wastewater Treatment Construction Grants Amendments	P.L. 97-117
1987	Water Quality Act of 1987	P.L. 100-4

Authorizations for appropriations to support the law generally expired at the end of FY1990 (September 30, 1990). Programs did not lapse, however, and Congress has continued to appropriate funds to carry out the act. Since the 1987 amendments, although Congress has enacted several bills that reauthorize and modify a number of individual provisions in the law, none comprehensively addressed major programs or requirements.

BACKGROUND

The Federal Water Pollution Control Act of 1948 was the first comprehensive statement of federal interest in clean water programs, and it specifically provided state and local governments with technical assistance funds to address water pollution problems, including research. Water pollution was viewed as primarily a state and local problem, hence, there were no federally required goals, objectives, limits, or even guidelines. When it came to enforcement, federal involvement was strictly limited to matters involving interstate waters and only with the consent of the state in which the pollution originated.

During the latter half of the 1950s and well into the 1960s, water pollution control programs were shaped by four laws that amended the 1948 statute. They dealt largely with federal assistance to municipal dischargers and with federal enforcement programs for all dischargers. During this period, the federal role and federal jurisdiction were gradually extended to include navigable intrastate, as well as interstate, waters. Water quality standards became a feature of the law in 1965, requiring states to set standards for interstate waters that would be used to determine actual pollution levels and control requirements. By the late 1960s, there was a widespread perception that existing enforcement procedures were too time-consuming and that the water quality standards approach was flawed because of difficulties in linking a particular discharger to violations of stream quality standards. Additionally, there was mounting frustration over the slow pace of pollution cleanup efforts and a suspicion that control technologies were being developed but not applied to the problems. These perceptions and frustrations, along with increased public interest in environmental protection, set the stage for the 1972 amendments.

The 1972 statute did not continue the basic components of previous laws as much as it set up new ones. It set optimistic and ambitious goals, required all municipal and industrial wastewater to be treated before being discharged into waterways, increased federal assistance for municipal treatment plant construction, strengthened and streamlined enforcement, and expanded the federal role while retaining the responsibility of states for day-to-day implementation of the law.

The 1972 legislation declared as its objective the restoration and maintenance of the chemical, physical, and biological integrity of the nation's waters. Two goals also were established: zero discharge of pollutants by 1985 and, as an interim goal and where possible, water quality that is both "fishable" and "swimmable" by mid-1983. While those dates have passed, the goals remain, and efforts to attain them continue.

OVERVIEW

The Clean Water Act (CWA) today consists of two parts, broadly speaking, one being the Title II and Title VI provisions, which authorize federal financial assistance for municipal sewage treatment plant construction. The other is regulatory requirements, found throughout the act, that apply to industrial and municipal dischargers.

The act has been termed a technology-forcing statute because of the rigorous demands placed on those who are regulated by it to achieve higher and higher levels of pollution abatement. Industries were given until July 1, 1977, to install "best practicable control technology" (BPT) to clean up waste discharges. Municipal wastewater treatment plants were required to meet an equivalent goal, termed "secondary treatment," by that date. (Municipalities unable to achieve secondary treatment by that date were allowed to apply for case-by-case extensions up to July 1, 1988. According to EPA, 86% of all cities met the 1988 deadline; the remainder were put under administrative or court-ordered schedules requiring compliance as soon as possible. However, many cities continue to make investments in building or upgrading facilities needed to achieve secondary treatment, and funding needs remain high; see discussion below.) Cities that discharge wastes into marine waters were eligible for case-by-case waivers of the secondary treatment requirement, where sufficient showing could be made that natural factors provide significant elimination of traditional forms of pollution and that both balanced populations of fish, shellfish, and wildlife and water quality standards would be protected.

The primary focus of BPT was on controlling discharges of conventional pollutants, such as suspended solids, biochemical oxygen demanding materials, fecal coliform and bacteria, and pH. These pollutants are substances that are biodegradable (i.e., bacteria can break them down), occur naturally in the aquatic environment, and deplete the dissolved oxygen concentration in water, which is necessary for fish and other aquatic life.

The act also mandated greater pollutant cleanup than BPT by no later than March 31, 1989, generally requiring that industry use the "best available technology" (BAT) that is economically achievable. BAT level controls generally focus on toxic substances. Compliance extensions of as long as two years are available for industrial sources utilizing innovative or alternative technology. Failure to meet statutory deadlines could lead to enforcement action.

The CWA utilizes both water quality standards and technology-based effluent limitations to protect water quality. Technology-based effluent limitations are specific numerical limitations established by EPA and placed on certain pollutants from certain sources. They are applied to industrial and municipal sources through numerical effluent limitations in discharge permits issued by states or EPA (see discussion of "Permits, Regulations, and Enforcement," below). Water quality standards are standards for the overall quality of water. They consist of the designated beneficial use or uses of a waterbody (recreation, water supply, industrial, or other), plus a numerical or narrative statement identifying maximum concentrations of various pollutants that would not interfere with the designated use. The act requires each state to establish water quality standards for all bodies of water in the state. These standards serve as the backup to federally set technology-based requirements by indicating where additional pollutant controls are needed to achieve the overall goals of the act. In waters where industrial and municipal sources have achieved technology-based effluent limitations, yet water quality standards have not been met, dischargers may be required to meet additional pollution control requirements. For each of these waters, the act requires states to set a total maximum daily load (TMDL) of pollutants at a level that ensures that applicable water quality standards can be attained and maintained. A TMDL is both a planning process for attaining water quality standards and a quantitative assessment of pollution problems, sources, and pollutant reductions needed to restore and protect a river, stream, or lake. Based on state reports, EPA estimates that more than 41,000 U.S. waters are impaired and require preparation of TMDLs.

Control of toxic pollutant discharges has been a key focus of water quality programs. In addition to the BPT and BAT national standards, states are required to implement control strategies for waters expected to remain polluted by toxic chemicals even after industrial dischargers have installed the best available cleanup technologies required under the law. Development of management programs for these post-BAT pollutant problems was a prominent element in the 1987 amendments and is a key continuing aspect of CWA implementation.

Prior to the 1987 amendments, programs in the Clean Water Act were primarily directed at point source pollution, wastes discharged from discrete and identifiable sources, such as pipes and other outfalls. In contrast, except for general planning activities, little attention had been given to nonpoint source pollution (runoff of stormwater or snowmelt from agricultural lands, forests, construction sites, and urban areas), despite estimates that it represents more than 50% of the nation's remaining water pollution problems. As it travels across land surface towards rivers and streams, rainfall and snowmelt runoff picks up pollutants, including sediments, toxic materials, and conventional wastes (e.g., nutrients) that can degrade water quality.

The 1987 amendments authorized measures to address such pollution by directing states to develop and implement nonpoint pollution management programs (Section 319 of the act). States were encouraged to pursue groundwater protection activities as part of their overall nonpoint pollution control efforts. Federal financial assistance was authorized to support demonstration projects and actual control activities. These grants may cover up to 60% of program implementation costs.

The CWA provides for special regulation of the discharge of oil or hazardous substances, because of the potentially catastrophic effects of such events on public health and welfare. Section 311 prohibits the discharge of oil or hazardous substances into U.S. waters. It also requires higher standards of care in the management and movement of oil, including a requirement for spill prevention plans; it enables the government to recover the costs of cleaning up oil and hazardous substance discharges; and it provides for penalties for such discharges. In 1990, Congress enacted the Oil Pollution Act, which partially amended Section 311 and established a comprehensive system for the cleanup of oil spills, adding a mechanism to impose liability for such spills.[2]

While the act imposes great technological demands, it also recognizes the need for comprehensive research on water quality problems. This is provided throughout the statute, on topics including pollution in the Great Lakes and Chesapeake Bay, in-place toxic pollutants in harbors and navigable waterways, and water pollution resulting from mine drainage. The act also authorizes support to train personnel who operate and maintain wastewater treatment facilities.

Federal and State Responsibilities

Under this act, federal jurisdiction is broad, particularly regarding establishment of national standards or effluent limitations. The EPA issues regulations containing the BPT and BAT effluent standards applicable to categories of industrial sources (such as iron and steel manufacturing, organic chemical manufacturing, petroleum refining, and others). Certain responsibilities can be assumed by qualified states, in lieu of EPA, and this act, like other environmental laws, embodies a philosophy of federal-state partnership in which the federal government sets the agenda and standards for pollution abatement, while states carry out day-to-day activities of implementation and enforcement. Responsibilities under the act that may be carried out by qualified states include authority to issue discharge permits to industries and municipalities and to enforce permits; 46 states have been authorized to administer this permit program. EPA issues discharge permits in the remaining states—Idaho, Massachusetts, New Hampshire, New Mexico— the District of Columbia, and most U.S. territories. In addition, as noted above, all states are responsible for establishing water quality standards.

TITLES II AND VI—MUNICIPAL WASTEWATER TREATMENT CONSTRUCTION

Federal law has authorized grants for planning, design, and construction of municipal sewage treatment facilities since 1956 (Act of July 9, 1956, or P.L. 84-660). Congress greatly expanded this grant program in 1972 in order to assist cities in meeting the act's pollution control requirements. Since that time Congress has authorized $65 billion and appropriated more than $88 billion in CWA funds to aid wastewater treatment plant construction. Grants are allocated among the states according to a complex statutory formula that combines two factors: state population and an estimate of municipal sewage treatment funding needs derived from a survey conducted periodically by EPA and the states. The most recent estimate indicated that, as of January 2008, $298 billion more would be required to build and upgrade municipal wastewater treatment plants in the United States and for other types of water quality improvement projects that are eligible for funding under the act, a 17% increase over the previous estimate from four years earlier.

Under the Title II construction grants program established in 1972, federal grants were made for several types of projects based on a priority list established by the states. Grants were generally available for as much as 55% of total project costs. For projects using innovative or alternative technology (such as reuse or recycling of water), as much as 75% federal funding was allowed. Recipients were responsible for non-federal costs but were not required to repay federal grants.

Policymakers have debated the balance between assisting municipal funding needs, which remain large, and the impact of aid programs such as the Clean Water Act's on federal spending and budget deficits. In the 1987 amendments, Congress balanced these competing priorities by extending federal aid for wastewater treatment construction through FY1994, yet providing a transition towards full state and local government responsibility for financing after that date. Grants under the previous Title II program were authorized through FY1990. Under Title VI of the act, grants to capitalize State Water Pollution Control Revolving Funds, or loan programs, were authorized beginning in FY1989 to replace the Title II grants. States contribute matching funds, and under the revolving loan fund concept, monies used for wastewater treatment construction will be repaid to a state fund, to be available for future construction in other communities.

All states now have functioning loan programs, but the shift from federal grants to loans has been easier for some than others. The new financing requirements have been a problem for cities (especially small towns) that have difficulty repaying project loans. Statutory authorization for grants to capitalize state loan programs expired in 1994; however, Congress has continued to provide annual appropriations. An issue affecting some cities is overflow discharges of inadequately treated wastes from municipal sewers and how cities will pay for costly remediation projects. In 2000, Congress amended the act to authorize a two-year $1.5 billion grant program to help cities reduce these wet weather flows. Authorization for that wet weather grant program expired at the end of FY2003 and has not been renewed.

PERMITS, REGULATIONS, AND ENFORCEMENT

To achieve its objectives, the CWA embodies the concept that all discharges into the nation's waters are unlawful, unless specifically authorized by a permit. Thus, more than 65,000 conventional industrial and municipal dischargers must obtain permits from EPA (or qualified states) under the act's National Pollutant Discharge Elimination System (NPDES) program (authorized in Section 402 of the act). NPDES permits also are required for more than 150,000 industrial and municipal sources of stormwater discharges. An NPDES permit requires the discharger (source) to attain technology-based effluent limits (BPT or BAT for industry, secondary treatment for municipalities, or more stringent where needed for water quality protection). Permits specify the effluent limitations a discharger must meet, and the deadline for compliance. Sources also are required to maintain records and to carry out effluent monitoring activities. Permits are issued for no more than five years and must be renewed thereafter to allow continued discharge.

The NPDES permit incorporates numerical effluent limitations issued by EPA. The initial BPT limitations focused on regulating discharges of conventional pollutants, such as bacteria and oxygen-consuming materials. The more stringent BAT limitations emphasize controlling toxic pollutants—

heavy metals, pesticides, and other organic chemicals. In addition to these limitations applicable to categories of industry, EPA has issued water quality criteria for more than 115 pollutants, including 65 named classes or categories of toxic chemicals, or "priority pollutants." These criteria recommend ambient, or overall, concentration levels for the pollutants and provide guidance to states for establishing water quality standards that will achieve the goals of the act.

A separate type of permit is required to dispose of dredged or fill material in the nation's waters, including wetlands. Authorized by Section 404 of the act, this permit program is administered by the U.S. Army Corps of Engineers, subject to and using EPA's environmental guidance. Some types of activities are exempt from permit requirements, including certain farming, ranching, and forestry practices which do not alter the use or character of the land; some construction and maintenance; and activities already regulated by states under other provisions of the act. EPA may delegate certain Section 404 permitting responsibility to qualified states and has done so twice (Michigan and New Jersey). For some time, the act's wetlands permit program has been one of the most controversial parts of the law. Some who wish to develop wetlands maintain that federal regulation intrudes on and impedes private land-use decisions, while environmentalists seek more protection for remaining wetlands and limits on activities that are authorized to take place in wetlands.

Nonpoint sources of pollution, which EPA and states believe are responsible for the majority of water quality impairments in the nation, are not subject to CWA permits or other regulatory requirements under federal law. They are covered by state programs for the management of runoff, under Section 319 of the act.

Other EPA regulations under the CWA include guidelines on using and disposing of sewage sludge and guidelines for discharging pollutants from land-based sources into the ocean. (A related law, the Ocean Dumping Act, 33 U.S.C. §§1401-45, regulates the intentional disposal of wastes into ocean waters.[3]) EPA also provides guidance on technologies that will achieve BPT, BAT, and other effluent limitations.

The NPDES permit, containing effluent limitations on what may be discharged by a source, is the act's principal enforcement tool. EPA may issue a compliance order or bring a civil suit in U.S. district court against persons who violate the terms of a permit. The penalty for such a violation can be as much as $25,000 per day. Stiffer penalties are authorized for criminal violations of the act—for negligent or knowing violations—of as much as $50,000 per day, three years' imprisonment, or both. A fine of as much as

$250,000, 15 years in prison, or both, is authorized for "knowing endangerment"—violations that knowingly place another person in imminent danger of death or serious bodily injury. Finally, EPA is authorized to assess civil penalties administratively for certain well-documented violations of the law. These civil and criminal enforcement provisions are contained in Section 309 of the act. EPA, working with the Army Corps of Engineers, also has responsibility for enforcing against entities who fail to obtain or comply with a Section 404 permit.

While the CWA addresses federal enforcement, the majority of actions taken to enforce the law are undertaken by states, both because states issue the majority of permits to dischargers and because the federal government lacks the resources for day-to-day monitoring and enforcement. Like most other federal environmental laws, CWA enforcement is shared by EPA and states, with states having primary responsibility. However, EPA has oversight of state enforcement and retains the right to bring a direct action where it believes that a state has failed to take timely and appropriate action or where a state or local agency requests EPA involvement. Finally, the federal government acts to enforce against criminal violations of the federal law.

In addition, individuals may bring a citizen suit in U.S. district court against persons who violate a prescribed effluent standard or limitation or permit requirement. Citizens also may bring suit against the Administrator of EPA for failure to carry out a nondiscretionary duty under the act.

SELECTED REFERENCES

Hamilton, Pixie, Timothy Miller, and Donna Myers. "Water Quality in the Nation's Streams and Aquifers—Overview of Selected Findings, 1991-2001." U.S. Geological Survey Circular 1265. May 2004.

Lavelle, Marianne. "Water Woes." U.S. News & World Report. June 4, 2007. pp. 37-53.

U.S. Congressional Budget Office. Public Spending on Transportation and Water Infrastructure. November 2010. Pub. No. 4088. 49 p.

U.S. Environmental Protection Agency. EPA's Draft Report on the Environment, Technical Document. Chapter 2, Purer Water. June 2003. EPA 600-R-03-050. pp. 2-1 - 2-64.

———. Clean Watersheds Needs Survey 2008, Report to Congress. May 2010. EPA-832-R-10-002.

——— Clean Water State Revolving Fund Programs 2009 Annual Report. June 2010. EPA-832-R- 10-001. 36 p.
——— National Water Quality Inventory: Report to Congress for the 2004 Reporting Cycle. January 2009. EPA-841-R-08-001. 37 p.
U.S. Government Accountability Office. Water Quality, A Catalog of Related Federal Programs. GAO/RCED-96-173. June 1996. 64 p.
——— Water Infrastructure: Information on Financing, Capital Planning, and Privatization. GAO-02-764. August 2002. 79 p.

Table 2. Major U.S. Code Sections of the Clean Water Act (codified generally as 33 U.S.C., Chapter 26, Sections 1251-1387)

33 U.S.C.	Section Title	Clean Water Act (as amended)
Subchapter I -	Research and Related Programs	
1251	Declaration of goals and policy	Sec. 101
1252	Comprehensive programs for water pollution control	Sec. 102
1253	Interstate cooperation and uniform laws	Sec. 103
1254	Research, investigations, training and information	Sec. 104
1255	Grants for research and development	Sec. 105
1256	Grants for pollution control programs	Sec. 106
1257	Mine water pollution control demonstrations	Sec. 107
1258	Pollution control in the Great Lakes	Sec. 108
1259	Training grants and contracts	Sec. 109
1260	Applications for training grants or contracts; allocations of grants or contracts	Sec. 110
1261	Award of scholarships	Sec. 111
1262	Definitions and authorizations	Sec. 112
1263	Alaska village demonstration project	Sec. 113
1264	Lake Tahoe study	Sec. 114
1265	In-place toxic pollutants	Sec. 115
1266	Hudson River PCB reclamation demonstration project	Sec. 116
1267	Chesapeake Bay	Sec. 117
1268	Great Lakes	Sec. 118
1269	Long Island Sound	Sec. 119
1270	Lake Champlain Basin program	Sec. 120
1273	Lake Pontchartrain Basin	Sec. 121

Clean Water Act: A Summary of the Law

33 U.S.C.	Section Title	Clean Water Act (as amended)
1274	Wet weather watershed pilot projects	Sec. 122
Subchapter II -	Grants for Construction of Treatment Works	
1281	Purpose	Sec. 201
1282	Federal share	Sec. 202
1283	Plans, specifications, estimates, and payments	Sec. 203
1284	Limitations and conditions	Sec. 204
1285	Allotment	Sec. 205
1286	Reimbursement and advanced construction	Sec. 206
1287	Authorization	Sec. 207
1288	Areawide waste treatment management	Sec. 208
1289	Basin planning	Sec. 209
1290	Annual survey	Sec. 210
1291	Sewage collection systems	Sec. 211
1292	Definitions	Sec. 212
1293	Loan guarantees for construction of treatment works	Sec. 213
1294	Public information on water recycling, reuse	Sec. 214
1295	Requirements for American materials	Sec. 215
1296	Determination of priority	Sec. 216
1297	Cost-effectiveness guidelines	Sec. 217
1298	Cost effectiveness	Sec. 218
1299	State certification of projects	Sec. 219
1300	Pilot program for alternative water source projects	Sec. 220
1301	Sewer overflow control grants	Sec. 221
Subchapter III -	Standards and Enforcement	
1311	Effluent Limitations	Sec. 301
1312	Water quality related effluent limitations	Sec. 302
1313	Water quality standards and implementation plans	Sec. 303
1314	Information and guidelines	Sec. 304
1315	Water quality inventory	Sec. 305
1316	National standards of performance	Sec. 306
1317	Toxic and pretreatment effluent standards	Sec. 307
1318	Inspections, monitoring, and entry	Sec. 308
1319	Federal enforcement	Sec. 309

Table 2. (Continued)

33 U.S.C.	Section Title	Clean Water Act (as amended)
1320	International pollution abatement	Sec. 310
1321	Oil and hazardous substance liability	Sec. 311
1322	Marine sanitation devices	Sec. 312
1323	Federal facilities pollution control	Sec. 313
1324	Clean lakes	Sec. 314
1325	National study commission	Sec. 315
1326	Thermal discharges	Sec. 316
1327	Financing study	Sec. 317
1328	Aquaculture	Sec. 318
1329	Nonpoint source management programs	Sec. 319
1330	National estuary program	Sec. 320
Subchapter IV -	Permits and Licenses	
1341	Certification	Sec. 401
1342	National pollutant discharge elimination system	Sec. 402
1343	Ocean discharge criteria	Sec. 403
1344	Permits for dredged or fill material	Sec. 404
1345	Disposal of sewage sludge	Sec. 405
1346	Coastal recreation water quality monitoring and notification	Sec. 406
Subchapter V -	General Provisions	
1361	Administration	Sec. 501
1362	General definitions	Sec. 502
1363	Water pollution control advisory board	Sec. 503
1364	Emergency powers	Sec. 504
1365	Citizen suits	Sec. 505
1366	Appearance	Sec. 506
1367	Employee protection	Sec. 507
1368	Federal procurement	Sec. 508
1369	Administrative procedure and judicial review	Sec. 509
1370	State authority	Sec. 510
1371	Other affected authority	Sec. 511
1372	Separability	Sec. 512
1372	Labor standards	Sec. 513
1373	Public health agency coordination	Sec. 514
1374	Effluent standards and water quality information advisory committee	Sec. 515

33 U.S.C.	Section Title	Clean Water Act (as amended)
1375	Reports to Congress	Sec. 516
1376	General authorization	Sec. 517
1377	Indian tribes	Sec. 518
1251 note	Short title	Sec. 519
Subchapter VI -	State Water Pollution Control Revolving Funds	
1381	Grants to states for establishment of revolving funds	Sec. 601
1382	Capitalization grant agreements	Sec. 602
1383	Water pollution control revolving loan funds	Sec. 603
1384	Allotment of funds	Sec. 604
1385	Corrective action	Sec. 605
1386	Audits, reports, and fiscal controls; intended use plan	Sec. 606
1387	Authorization of appropriations	Sec. 607

Note: This table shows only the major code sections. For more detail and to determine when a section was added, the reader should consult the official printed version of the U.S. Code.

End Notes

[1] For example, see CRS Report R41594, Water Quality Issues in the 112th Congress: Oversight and Implementation, by Claudia Copeland.

[2] P.L. 101-380; 33 U.S.C. § 2701 et seq.

[3] CRS Report RS20028, Ocean Dumping Act: A Summary of the Law, by Claudia Copeland.

INDEX

#

20th century, 43
21st century, 9

A

abatement, xi, 93, 94, 97, 99, 106
accounting, 10
adaptability, 36
adaptation, 9
adverse effects, 15
advocacy, 60
aesthetic, 6, 62
agencies, ix, 4, 6, 11, 12, 13, 16, 18, 19, 22, 27, 28, 30, 31, 34, 35, 42, 43, 51, 52, 53, 55, 56, 57, 58, 59, 60, 61, 64, 66, 71, 72, 79, 86, 90
agency actions, 52
aggregation, 58
agricultural market, 9
agriculture, 8, 26, 35, 61, 66
Alaska, 7, 23, 63, 86, 88, 104
appropriations, 3, 22, 28, 60, 95, 101, 107
aquatic life, 12
assessment, 30, 31, 98
atmosphere, 10
Attorney General, 72
authority(s), ix, x, 12, 13, 14, 15, 16, 19, 20, 21, 22, 30, 31, 42, 43, 47, 49, 51, 56, 61, 66, 67, 68, 69, 77, 78, 79, 80, 81, 84, 88, 89, 90, 91, 99, 106

B

bacteria, xi, 94, 97, 101
banking, 25, 35
banks, 35, 39, 85
barriers, 32, 71
benefits, 3, 10, 25, 26, 28, 59
biomass, 10
birds, x, 6, 16, 45, 62, 63, 78, 79, 80, 83, 86, 87
bogs, vii, 2, 44, 85, 86
breakdown, 54
budget deficit, 100

C

canals, ix, 41
carbon, 10, 36
carbon dioxide, 10
category b, 23
certification, 61, 67, 84, 88, 91, 105
CFR, 73, 90
challenges, 19, 22, 48, 59, 69
chemical(s), 49, 53, 58, 85, 96, 98, 99, 102
Chicago, 79
Chief Justice, 71, 72, 81
city(s), 72, 79, 97, 100, 101

clarity, 18, 63
classes, 102
classification, 23, 73
Clean Water Act (CWA), v, vii, ix, x, xi, 1, 3, 4, 5, 10, 11, 17, 27, 36, 37, 38, 41, 43, 48, 54, 60, 63, 71, 72, 73, 74, 77, 78, 93, 94, 95, 96, 98, 100, 104, 106
cleaning, 99
cleanup, 33, 84, 96, 97, 98, 99
climate, 9, 10
climate change, 10
Clinton Administration, 8, 12
CO_2, 10, 36
coal, 12, 13, 14, 15
coastal areas, vii, 3, 7, 33, 44
coatings, 33
collaboration, 8
commerce, 43, 45, 51, 68, 73, 80, 82, 83, 86, 87
commercial, 7, 26, 83
commodity, 23
community(s), x, xii, 5, 24, 42, 54, 57, 62, 63, 94, 101
compensation, 29, 35, 67
competing interests, 64
complexity, viii, 2
compliance, 25, 71, 97, 101, 102
conditioning, 67
conference, 43, 44, 55, 80
conflict, 85
Congressional Budget Office, 103
connectivity, 65
consensus, 9
consent, 96
conservation, xi, 4, 8, 24, 25, 26, 27, 28, 31, 66, 78
conservation programs, 4, 24, 25, 26
conserving, 30
consolidation, 46
constant rate, 32
Constitution, 21, 29, 49, 52, 80, 90
construction, xi, 32, 35, 36, 64, 93, 94, 96, 97, 98, 100, 102, 105
control measures, 61

controversial, ix, 3, 12, 14, 15, 22, 23, 24, 42, 79, 102
controversies, viii, 2, 6, 35
convention, 29
cooperation, 104
coordination, 4, 12, 30, 106
cost, 13, 25, 26, 28, 32
counsel, 83, 91
Court of Appeals, 80
covering, 8, 84
criticism, 14, 27, 35, 53, 60, 69
crop, 25
crop insurance, 25
crop production, 25
CRP, 26
cycles, 44

D

damages, 7
danger, 103
database, 64
decision control, 48
degradation, 10, 45, 73, 80, 84, 87
demonstrations, 104
denial, 16
Department of Agriculture, 6, 68
Department of Defense, 36, 73
Department of Justice, 72
Department of the Interior, 68
destruction, vii, 3, 34, 36, 45, 73, 80, 84, 87
differential treatment, 23
direct action, 103
direct payment, 3
directives, 89
disappointment, 47, 54
discharges, vii, xi, 1, 7, 11, 12, 15, 20, 48, 61, 68, 69, 79, 87, 91, 93, 94, 97, 98, 99, 101, 106
dissolved oxygen, 97
District of Columbia, 100
divergence, 51
draft, viii, ix, 2, 19, 20, 22, 42, 56, 60, 64, 65, 74
drainage, ix, 10, 41, 47, 69, 99

drinking water, 64

E

economic activity, 82
ecosystem, 9, 10, 62
education, 6, 31
effluent, xii, 94, 97, 99, 101, 102, 103, 105
emission, 10
endangered, 6, 61
endangered species, 6, 61
energy, 9
enforcement, xii, 24, 52, 54, 55, 61, 88, 94, 96, 97, 99, 102, 103, 105
enlargement, 43
enrollment, 4, 26
environment(s), xi, 15, 33, 67, 82, 84, 94, 97
environmental effects, 11
environmental impact, 14, 23
environmental issues, 12
environmental protection, 83, 90, 96
Environmental Protection Agency, vii, 1, 4, 36, 37, 38, 43, 67, 72, 73, 74, 79, 103
environmental services, viii, 2
erosion, 8, 10, 23, 31, 83
evolution, 11
exclusion, 55
Executive Order, 87
exercise, 67
expertise, 44

F

faith, 25
Farm Bill, 37
farm land, xi, 94
farm programs, vii, 3, 24
farmers, 3, 4, 24
farms, 26
fear, 23, 27
federal agency, 12, 87
federal aid, 100
federal assistance, 96
federal courts, 4, 16, 43
federal funds, 26
federal government, xii, 28, 34, 43, 44, 87, 88, 94, 99, 103
federal law, vii, x, 3, 5, 42, 62, 66, 102, 103
Federal Register, 36, 37, 38, 73, 74
federal regulations, vii, 1
federal tax code, vii, 3
federalism, 49, 82
fens, vii, 2, 44
Fifth Amendment, 50
filters, 85
filtration, 49
financial, xi, 10, 20, 24, 25, 89, 93, 97, 99
financial resources, 20
financial support, 89
fish, 6, 7, 11, 13, 28, 32, 33, 97
Fish and Wildlife Service, viii, 2, 6, 7, 28, 36, 38, 86, 90
flexibility, 23, 25
flight, 83
flooding, vii, 3, 33, 44, 84
floods, 6, 57, 62
food, 6, 33, 62, 85
food chain, 33
food production, 6, 62
Food, Conservation, and Energy Act of 2008, 4, 24
football, 32
force, 19, 32
formula, 100
freshwater, 9, 32, 66, 88
funding, 22, 26, 27, 28, 32, 60, 97, 100
funds, xii, 33, 94, 95, 100, 101, 107

G

General Accounting Office (GAO), 17, 32, 34, 35, 37, 38, 53, 73, 104
Georgia, 88
GHG, 10, 36
governments, 4
grants, xii, 28, 94, 99, 100, 101, 104, 105
grasses, 33
grasslands, 26

Index

greenhouse, 10
greenhouse gas (GHG), 10
groundwater, 31, 85, 99
growth, 13
guidance, viii, ix, 2, 3, 6, 17, 18, 19, 20, 21, 22, 23, 28, 35, 42, 43, 51, 52, 53, 54, 55, 56, 57, 58, 59, 60, 61, 63, 67, 68, 69, 71, 72, 73, 84, 89, 102
guidelines, 13, 31, 60, 90, 95, 102, 105
Gulf Coast, 33
Gulf of Mexico, 8, 32, 33, 34

H

habitat(s), vii, viii, x, 2, 3, 6, 7, 9, 11, 26, 27, 28, 33, 42, 58, 62, 78, 80, 83
harbors, 99
Hawaii, 72
hazardous substances, xii, 94, 99
health, 106
heavy metals, 102
history, 15, 82
House, 21, 25, 27, 29, 37, 55, 68, 73, 74, 75
House Committee on Transportation and Infrastructure, 73
House of Representatives, 37, 74
human, 5, 7, 9, 31
hurricanes, 32, 38
hydrologic regime, 6

I

identification, 6
image, 23
impairments, 102
imprisonment, 102
improvements, 43, 59, 94
income, 24
income support, 24
indirect effect, 7
individuals, 29, 103
industry(s), xi, 14, 16, 54, 60, 64, 93, 94, 97, 100, 101, 102
infrastructure, 9
injury, 103
inspections, 55
integrity, x, 9, 42, 49, 53, 58, 62, 65, 85, 96
interest groups, 17, 63
interstate waters, x, 44, 45, 57, 78, 80, 83, 96
investments, 97
iron, 99
irrigation, 59
isolation, 58, 65
issues, vii, ix, xi, 1, 2, 3, 4, 10, 17, 18, 19, 20, 21, 29, 41, 46, 49, 51, 52, 54, 55, 60, 63, 67, 68, 71, 72, 78, 79, 82, 89, 95, 99

J

judicial interpretation, 79

L

Lake Pontchartrain, 104
lakes, vii, ix, 2, 20, 26, 31, 41, 44, 45, 48, 59, 65, 69, 80, 82, 86, 106
landscape, 58, 62
laws, vii, viii, xi, 2, 3, 13, 20, 28, 29, 35, 38, 61, 67, 78, 88, 96, 99, 103, 104
laws and regulations, 29
lead, 6, 23, 97
legal protection, 61
legislation, vii, viii, ix, xi, 1, 3, 4, 5, 14, 21, 22, 24, 25, 26, 27, 31, 32, 37, 42, 43, 44, 60, 66, 67, 68, 69, 75, 82, 88, 93, 94, 96
legislative authority, 21
legislative proposals, 24, 32
life cycle, 33
light, 16, 18, 22, 30, 50, 64, 69, 71, 84
litigation, 14, 18, 45, 51, 52, 83, 84
loans, 101
local authorities, xi, 78
local government, xi, 78, 87, 95, 100
Louisiana, 31, 32, 33, 34, 38, 88

M

majority, ix, 19, 31, 42, 45, 46, 47, 48, 50, 59, 81, 83, 102, 103
man, 29, 46, 47, 71
management, 10, 25, 27, 35, 87, 88, 98, 99, 102, 105, 106
manufacturing, 14, 99
mapping, 88
marsh, 33
marshes, vii, 2, 32, 33, 36, 44, 82, 85
materials, x, 3, 11, 18, 36, 65, 77, 78, 82, 97, 98, 101, 105
matter, 16, 24, 46, 49, 80, 84
metals, 102
Mexico, 28, 100
Miami, 37
migratory waterfowl, vii, 3
military, 36
Mississippi River, 32
moderates, 50
modifications, 15, 21, 25, 57
modules, 73
mortality, 33
mosaic, 58

N

National Academy of Sciences, 38
national policy, viii, 2
National Research Council, 34, 38, 85, 90
natural disaster, 7
natural resources, 88
Natural Resources Conservation Service (NRCS), 6
navigable waters, ix, x, 18, 19, 41, 42, 43, 44, 45, 46, 47, 49, 52, 53, 54, 55, 57, 58, 61, 62, 66, 68, 71, 73, 77, 78, 79, 80, 81, 82, 83, 84, 85, 86, 87, 89, 91
NOAA, 7
North America, 3, 28
NRC, 34, 35, 85
NRCS, 11, 25, 27
nutrients, 85, 98

O

Obama, viii, 1, 3, 19, 21, 55, 68, 74
Obama Administration, viii, 1, 3, 19, 21, 55, 68
objectivity, 6
oceans, 65
Office of Management and Budget, viii, ix, 2, 19, 42, 56, 64
officials, 14, 21, 23, 36, 56, 60, 62, 64, 66, 68
oil, xii, 8, 33, 61, 84, 87, 94, 99
Oil Pollution Act, 61
oil spill, 33, 61, 84, 87, 99
OMB, 19, 56, 64, 65
operations, 14
opportunities, 32, 52
organic chemicals, 102
Outer Continental Shelf, 38
outreach, 31
oversight, 103
ownership, 66, 85
oxygen, 97, 101

P

Pacific, 72
participants, 54
peer review, 65
penalties, 25, 99, 102
permeation, 84
permit, vii, xii, 1, 3, 12, 13, 14, 15, 16, 17, 22, 23, 26, 27, 30, 34, 35, 36, 44, 49, 60, 61, 67, 79, 84, 87, 88, 90, 91, 94, 100, 101, 102, 103
petroleum, 99
pH, 97
physical characteristics, vii, 2, 3, 44, 85
plants, xii, 5, 6, 13, 33, 36, 62, 85, 87, 94, 97, 100
playa lakes, vii, 2, 20, 26, 44, 65, 69, 80
policy, vii, ix, 1, 2, 3, 5, 8, 10, 17, 34, 42, 56, 60, 61, 63, 65, 79, 85, 86, 104
policy makers, 5

policymakers, ix, 10, 24, 34, 42, 61, 89
politics, 64
pollutants, xi, 9, 53, 57, 58, 61, 94, 96, 97, 98, 99, 101, 102, 104
pollution, xi, 43, 57, 67, 80, 88, 93, 94, 95, 96, 97, 98, 99, 100, 102, 104, 106, 107
ponds, 9, 20, 31, 59, 69, 71, 79, 80, 81, 86
pools, 10, 65
population, 7, 100
population growth, 7
precedent(s), 72, 84
predictability, 24
preparation, 98
President, 8, 36, 74
President Clinton, 8
prevention, 61, 99
principles, viii, 1, 21, 68
private banks, 35
private sector, 8
procurement, 106
producers, 25, 26
program administration, 5
project, 14, 20, 28, 32, 35, 37, 67, 88, 100, 101, 104
property rights, 29, 50
protection, vii, viii, xi, 2, 3, 4, 5, 6, 11, 17, 19, 20, 23, 24, 25, 26, 27, 29, 30, 31, 34, 38, 54, 57, 60, 62, 66, 67, 78, 82, 85, 86, 87, 88, 89, 99, 101, 102, 106
public health, 99
public interest, 96

Q

quality improvement, xii
quality standards, 61, 96, 98, 105

R

rainfall, 98
Rapanos v. United States, ix, 4, 17, 41, 43, 72, 73, 74
reading, 16, 45, 48, 50, 51, 52, 63, 80, 81, 82, 86, 89, 90

reality, 22
recommendations, 35
recovery, 33
recreation, 6, 62, 86, 98, 106
recreational, 13
recreational areas, 13
recycling, 100
Reform, 73
regulations, viii, ix, 2, 4, 12, 19, 20, 22, 23, 24, 27, 37, 42, 44, 45, 46, 47, 49, 52, 56, 57, 63, 64, 66, 68, 69, 71, 73, 79, 80, 88, 99, 102
regulatory changes, 79
regulatory requirements, xi, 61, 93, 97, 102
reintroduction, 32
relative size, 73
relief, 38
remediation, 101
replacement ratios, 31
requirements, vii, 1, 3, 22, 27, 28, 31, 35, 50, 52, 60, 61, 87, 88, 95, 96, 98, 100, 101, 102
reserves, 67, 85
resolution, 22, 69
resources, xi, 9, 10, 14, 15, 20, 30, 31, 35, 57, 78, 86, 89, 103
response, vii, 1, 18, 23, 25, 32, 33, 35, 44, 69, 83, 89
restoration, vii, 1, 2, 3, 10, 25, 30, 31, 32, 34, 35, 96
restoration programs, 3
restrictions, 15, 29
rights, viii, 2, 29, 67
risk, 24, 64
roots, 10, 33
rule of law, 51
rules, ix, 3, 12, 16, 17, 18, 42, 48, 50, 64, 66, 67, 88
runoff, xi, 94, 98, 102
rural areas, 66
rural development, 8

S

SAB, 65

salinity, 36
salinity levels, 36
saltwater, 8, 31
savings, 69
school, 30, 82
science, viii, 2, 5, 56, 57, 64, 65
Science Advisory Board (SAB), 65
scientific knowledge, 36
scope, ix, x, 4, 16, 17, 18, 19, 21, 22, 41, 42, 51, 52, 56, 57, 59, 60, 62, 63, 66, 68, 69, 71, 72, 73, 77, 78, 82, 83, 84, 89
sea level, 7, 9, 31, 36
sea-level, 9
sea-level rise, 9
sediment(s), 9, 32, 33, 85, 98
semantics, 50
Senate, ix, 4, 18, 20, 21, 25, 27, 37, 42, 64, 68, 69, 72, 74
services, 7, 9, 79, 85
sewage, xi, 87, 93, 94, 97, 100, 102, 106
sewage treatment plant, xi, 93, 94, 97
shape, 83, 94
shellfish, 6, 7, 32, 97
shoots, 33
shoreline, 32, 33
shores, 84
showing, 97
Sierra Club, 72, 74
sludge, 102, 106
soil erosion, 33
soil type, 13
solid waste, 79
solution, 23
South Dakota, 88
species, 9, 13, 26, 58
specifications, 105
spending, 100
staffing, 89
stakeholder groups, 54
stakeholders, 20, 64, 66
statutes, 80, 81, 82, 88, 90, 91
statutory provisions, 22
steel, 99
stock, 59
storage, 6, 62
storms, 8, 9, 34
stormwater, 98, 101
subsidy, 3
Supreme Court, v, vii, ix, x, 1, 4, 14, 16, 17, 21, 27, 29, 37, 41, 43, 44, 46, 47, 48, 49, 50, 51, 56, 60, 61, 63, 68, 72, 73, 77, 78, 79, 80, 81, 82, 83, 84, 89, 90
surface mining, 14
SWANCC, v, vii, x, 1, 4, 6, 16, 17, 18, 19, 20, 21, 26, 27, 37, 45, 46, 47, 48, 51, 52, 56, 62, 63, 66, 68, 69, 71, 72, 73, 74, 77, 78, 79, 80, 81, 82, 83, 84, 85, 86, 87, 89, 91

T

tax incentive, 30
technical assistance, 28, 31, 89, 95
technology(s), xi, 93, 96, 97, 98, 100, 101, 102
territorial, 20, 69
territory, 28
threats, 12
tides, vii, 3, 33, 44
timber production, 6
Title I, 96, 100
Title II, 96, 100
Title V, 96, 101
total costs, 28
toxic pollutant discharges, xi, 94, 98
toxic substances, 97
toxicity, 33
training, 104
transparency, 53
transportation, 60
treatment, xi, 23, 57, 93, 94, 96, 97, 99, 100, 101, 105
tundra, 86

U

U.S. Army Corps of Engineers, x, 4, 16, 17, 32, 38, 43, 44, 45, 72, 77, 78, 102
U.S. Department of the Interior, 36, 38

U.S. Geological Survey (USGS), 31, 38, 103
uniform, 104
universities, 30
urban, xi, 7, 8, 85, 94, 98
urban areas, xi, 94, 98
urbanization, 64
USDA, 25, 26, 37
USGS, 32

V

vegetation, 13, 26, 32, 33, 85
vein, 17
veto, 12, 13, 14, 15, 36, 88, 90
violence, 82
vision, 82
vote, 47
voting, 51, 81
vulnerability, 8

W

Washington, 38
waste, vii, 1, 12, 14, 20, 69, 91, 97, 105
waste disposal, 14
waste treatment, 20, 69, 105
wastewater, xii, 94, 96, 97, 99, 100

water purification, 6, 62
water quality, x, xi, 11, 14, 27, 30, 49, 59, 61, 66, 67, 78, 79, 84, 85, 88, 93, 94, 96, 97, 98, 99, 100, 101, 102, 106
water quality standards, 30, 61, 67, 88, 96, 97, 100, 102
water resources, 27
water supplies, 13
watershed, 9, 58, 73, 88, 105
waterways, 26, 43, 96, 99
websites, 53
welfare, 99
wetland acreage, viii, x, 2, 7, 78, 86, 88
wetland areas, vii, 3, 10, 11, 13, 23, 34, 44
wetland mitigation, 36
wetland restoration, 10, 28, 32, 33, 38, 66
White House, 19, 64, 89
wildlife, viii, x, 2, 7, 11, 13, 14, 25, 26, 27, 28, 32, 33, 42, 62, 86, 97
Wisconsin, 20
witnesses, 18
World Bank, 10, 36
worldwide, 9
wrestling, 43

Y

yield, 32